Introduction to Plant Fossils

This book provides an excellent practical introduction to the study of plant fossils, and is especially written for those who have had little previous experience of this type of palaeontology. The illustrated text summarises the main groups of plants that occur as fossils and explains how best to investigate them. It provides useful guidance about modern research techniques that reveal hidden details of anatomical and reproductive characteristics, and of the kinds of features that are used to identify the most commonly found plant fossils. The main approaches for interpreting plant fossils are assessed, and the book highlights how such methods can be employed by the palaeobotanist to increase our knowledge of plant evolution, palaeoecology, palaeogeography and stratigraphy.

This guide to plant fossils includes a discussion on how the science of palaeobotany has developed over the last 300 years, and incorporates examples from a global range of plant groups. It is essential reading for students of introductory or intermediate courses in palaeobotany, palaeontology and plant evolution, and a valuable resource for amateurs looking for help in identifying and studying plant fossils.

- Extensive illustrations of plant fossils and living plants enable the reader to think of fossils as once-living organisms rather than parts of 'dead plants'
- Set-aside boxes describing the key characteristics of major groups of plant fossils provide expert guidance on their identification
- A chapter on the techniques available for the study of plant fossils assists the reader on how to approach the subject
- A chapter on the changing patterns of vegetation through time provides an overview of how and why vegetation has changed
- Emphasis on compressions and impressions, the most common types of plant fossil, is tailored to aid student understanding

Christopher J. Cleal is Head of Vegetation History Section at the National Museum Wales. He obtained his BSc and PhD from the University of Sheffield, and has studied Palaeozoic palaeobotany and stratigraphy for over 35 years, with special reference to the link between vegetation and climate change. He has worked at the University of Sheffield, the Museum of the Saarbrücken Mining School in Germany, the Nature Conservancy Council, and the National Museum of Wales. He is a Fellow of the Geological Society, London, and the Linnean Society, London, a former Council Member of the Palaeontological Association, and Secretary of the British Institute for Geological Conservation. He has published a number of books, and over 100 papers in academic journals, dealing with palaeobotany, stratigraphy and geoconservation. He was on the Editorial Board of the journal *Palaeontology*, has been co-editor of the biennial *Bibliography of European Palaeobotany and Palynology* for the last 10 years, and is currently an editor of *Systematic Palaeontology*.

Barry A. Thomas holds an honorary Chair at Aberystwyth University, Wales. He obtained his BSc from the University of Sheffield, and his Ph.D. and D.Sc. from the University of Reading. He has studied Palaeozoic palaeobotany and stratigraphy for over 45 years, with special reference to Carboniferous floras and pteridophytes. He was Head of the Life Sciences Department and Dean of Science at Goldsmiths' College, London, Keeper of the Botany Department, National Museum, Wales, and Professor in the Geography Department, Lampeter. He is a Fellow of the Linnean Society, London, member of the Palaeontological Association, Treasurer of the British Institute for Geological Conservation, and past President of the British Pteridological Society. He has published numerous books and over 120 papers in academic journals, dealing with pteridophytes, palaeobotany, stratigraphy and geoconservation. He started the biennial *Bibliography of European Palaeobotany and Palynology*, and is currently an editor of *Law Science and Policy*.

An Introduction to Plant Fossils

Christopher J. Cleal
Barry A. Thomas

CAMBRIDGE
UNIVERSITY PRESS

CAMBRIDGE UNIVERSITY PRESS
Cambridge, New York, Melbourne, Madrid, Cape Town, Singapore,
São Paulo, Delhi

Cambridge University Press
The Edinburgh Building, Cambridge CB2 8RU, UK

Published in the United States of America by Cambridge University Press,
New York

www.cambridge.org
Information on this title: www.cambridge.org/9780521887151

First published 2009

Printed in the United Kingdom at the University Press, Cambridge

A catalogue record for this publication is available from the British Library

ISBN 978-0-521-88715-1 hardback
ISBN 978-0-521-71512-6 paperback

Cambridge University Press has no responsibility for the persistence or
accuracy of URLs for external or third-party Internet websites referred to
in this publication, and does not guarantee that any content on such
websites is, or will remain, accurate or appropriate.

Contents

Preface

This book introduces the reader to the study of plant fossils, and is especially aimed at those who have had little previous experience of this type of palaeontology. We have provided an illustrated text that summarises the main groups of plants that occur as fossils and explains how one should go about investigating them. Practical guidance is given on techniques that can be used to reveal hidden details of anatomical and reproductive characters, as well as the sorts of features that are used to identify some of the most commonly found plant fossils. Many of the approaches that are used to interpret plant fossils are discussed and we show how they can be used by the palaeobotanist to increase our knowledge of plant evolution, palaeoecology, palaeogeography and stratigraphy.

Although we have covered the main groups of fossils that the reader is likely to encounter, there has inevitably been some subjectivity in the coverage, reflecting our own research interests that are focussed on Late Palaeozoic floras of Euramerica and China. Nevertheless, we trust that the sorts of approaches we have given will have a wide applicability, and will be of help to all who are looking at plant fossils for the first time.

Many of the illustrations that we have included here were originally prepared for our previous book entitled *Plant Fossils*, our field guide on Coal Measures plants published by the Palaeontological Association, and the two volumes in the Geological Conservation Review series that we co-authored. The drawings are mostly by Annette Townsend, Deborah Spillards and Dale Evans of Amgueddfa Cymru – National Museum Wales (hereafter referred to as NMW) to whom we are deeply indebted. Others are based on paintings by Annette Townsend and the late Pauline Dean, prepared for books that we published on past plant life (Pardoe & Thomas, 1992; Thomas & Cleal, 1993, 1998, 2000); these paintings are in the collections of NMW. We are also indebted to many friends and colleagues who have made available photographs that illustrate this book: John Anderson (Pretoria), Sidney Ash (Albuquerque), Richard Bateman (Birmingham), Bill Chaloner (London), Margaret Collinson (London), Bill DiMichele (Washington DC), Tatyana Dimitrova (Sofia), Muriel Fairon-Demaret (Liège), Else Marie Friis (Stockholm), Pat Gensel (North Carolina), Alan Hemsley (Cardiff), Hans Kerp (Münster), Paul Kenrick (London), Jiři and Zlatko Kvaček (Prague), Jean-Pierre Laveine (Lille), Franz-Josef Lindemann (Oslo), Steve Manchester (Gainesville, Florida), Nick Rowe (Montpellier), Ruth Stockey (Edmonton), Don Tidwell (Utah), Maria Aquirre Urreta (Argentina), Bob Wagner (Cordoba), Lynda Warren (Aberystwyth), Joan Watson (Manchester), Wang Ziqiang (Tianjin), Erwin Zodrow (Sydney, Nova Scotia) and Nickolas Zouros (Lesvos). Many of the other photographs (notably those previously included in *Plant Fossils*) were taken by the Photography Department of NMW, to whom we are also very grateful. Thanks go to Charles Wellman (University of Sheffield) who first suggested we write this book, and to Lynda Warren whose careful reading of the manuscript made many improvements. Finally, we are extremely grateful to the staff of Cambridge University Press for their patience and help in seeing this book through to completion.

Chapter 1

Introduction

The popular image of a 'typical' fossil is often the remains of an animal – a seashell or a dinosaur bone. Animals, especially invertebrates, do indeed tend to be more commonly found as fossils, but plants can be fossilised given the right conditions. Plants probably first appeared on land more than 400 million years ago, during the Ordovician Period, and since that time their remains have found their way into muds, silts and sands, to be preserved as fossils (Fig. 1.1 shows the standard stratigraphical names for the time intervals referred to in this book). These plant fossils give us our only direct view of the vegetation of the past and from this we can infer the history of the evolution of plants and floras.

What is a plant?

This may seem obvious but, when you examine the problem in detail, it is not. Early naturalists divided living organisms into two kingdoms, animals and plants. Animals were thought of as mobile organisms whose nutrition was based on the consumption of other organisms, whereas plants were static, green organisms whose food was generated by internal processes (mainly photosynthesis). Although this was satisfactory for classifying most of the organisms that we meet on a day-to-day basis, it soon became evident that the distinction was not clear-cut. Fungi were the most obvious discrepancy because their nutrition is based on a saprophytic existence, or on the decomposition of plant and animal residues. Fungi were therefore

eventually assigned to their own kingdom. As biologists looked more closely at the microscopic world, the position became even more complex.

Many authors today do not regard plants as a systematically coherent group of organisms. The organisms that most of us would refer to as land plants are instead sometimes referred to as embryophytes, which are formally defined as those organisms that have alternating sexual (gametophyte) and asexual (sporophyte) generations, and where the gametophyte produces an embryo (alternating generations, Fig. 4.2). Some simple algal organisms have alternating sexual and asexual generations, but they do not produce embryos. Animals produce embryos but of a fundamentally different type, consisting of a hollow ball of cells that is usually detached from the tissue of the mother; the embryophyte embryo, in contrast, is a solid structure that remains embedded in the maternal tissue.

Whilst accepting this formal definition of embryophytes, for convenience we will continue through the rest of this book to refer to them as plants. In this context, plants consist of charophytes, bryophytes and vascular plants (also known as tracheophytes), together with some primitive non-vascular plants found mainly in Early Palaeozoic floras. Bryophytes, which today include mosses, hornworts and liverworts, are perfectly adapted to life on land but have never developed into large organisms, as have the vascular plants. Vascular plants include most land vegetation such as ferns, sphenophytes ('horsetails'), lycophytes ('club mosses'), gymnosperms and angiosperms. Their main defining feature is a

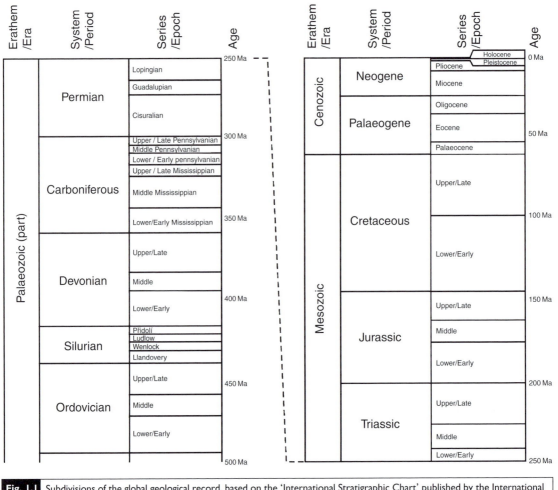

Fig. 1.1 Subdivisions of the global geological record, based on the 'International Stratigraphic Chart' published by the International Commission on Stratigraphy. The lowest Palaeozoic subdivision (the Cambrian System/Period) is not shown. The approximate dates given at the right of the two columns, based mainly on radiometric dating, are given in millions of years (denoted as Ma).

stem with vascular tissue, which assists in supporting the plant and in the transport of nutrients and water around its body; most vascular plants also have a dominant sporophyte phase (although some primitive vascular plants appear to have had sporophytes and gametophytes of similar size, see Chapter 4).

How do plant fragments get into the fossil record?

Very occasionally, whole plants became entrapped in sediment and preserved where they were growing; such fossils are referred to as autochthonous (examples are shown in Figs. 1.2 and 5.20). More usually, we find fragments (leaves, seed, branches, etc.) that have been detached from the plant. The process of fragmentation may be part of the natural strategy of the plant, such as the deciduous shedding of leaves in many angiosperm trees. More often, however, it was the result of traumatic external influences, such as storm damage.

The detached fragment may have fallen to the ground near where the parent plant was growing, in which case the fossil is described as hypautochthonous. Such situations tend to be rare because the ground surrounding a growing plant tends to be exposed and the tissue of the

Fig. 1.2 a. *Calamites* stem preserved *in situ* in a Middle Pennsylvanian fluvio-lacustrine sequence at Brymbo Quarry, near Wrexham, Wales. b. *In situ* silicified coniferous trees near Agua del Zorro, Mendoza Province, Argentina. The trees were discovered by Charles Darwin in 1835, while on his travels around the world aboard the Beagle and are described in two of his publications (Darwin, 1839, 1844). This Triassic site, now known as 'Darwin's Fossil Forest', is protected by Provincial and National laws. Photos by B. A. Thomas (a) and M. A. Urreta (b).

detached plant fragment will normally decay rapidly. Preservation is more likely when the plant remains are transported away from the place of growth usually by air (wind) or water (rivers), to where it can be rapidly covered by sediment, such as in a lake – these form allochthonous fossils.

Because of the effects of this fragmentation and transportation, most plant fossil assemblages normally represent a mixture of plant fragments from different habitats rather than a single natural assemblage of plants. Reconstructions of original vegetation can only be directly achieved with autochthonous or hypautochthonous assemblages, such as 'fossil forests' of tree stumps where the landscape has been subjected to a sudden catastrophic inundation by sediment. Most peat deposits are hypautochthonous and so, if their component plant remains can be identified such as in coal balls (discussed later in this chapter, and in Chapter 2), they can also often give a reasonably accurate picture of what the local vegetation looked like. The study of spores and pollen (palynology) in peat can be similarly useful here, provided the plants that produced the different types of pollen and spores are known – an assumption that is reasonable in Cenozoic and most Mesozoic floras, but can be problematic in Palaeozoic floras that include extinct and sometimes poorly understood plant groups.

A consequence of the fossilisation process is that most plant fossils are only fragments of the original plant. Even some of the larger organs, such as fronds of ferns or Palaeozoic gymnosperms, may have been fragmented. Reconstructions of plants depend on piecing together the chance finds of attached organs, such as a leaf and a seed, a leaf and a stem, and a stem and a cone. Mostly, we know nothing of the seeds that were borne by the plant that produced a particular leaf, let alone the appearance of the whole plant. This has consequences for naming plant fossils, which we will deal with in Chapter 3.

Types of plant fossil

When the plant fragment becomes trapped within the sediment, it undergoes various changes, which can affect what sort of detail is retained. Palaeobotanists have developed a range of terms to describe the different modes of preservation (Fig. 1.3).

Most plant fossils represent fragments of plant that have become trapped in sediment where anaerobic conditions prevent microbial breakdown of their tissues. They are then flattened by the weight of the overlying sediment, although retain their overall shape. Such fossils are known as adpressions (Fig. 1.4a). The plant

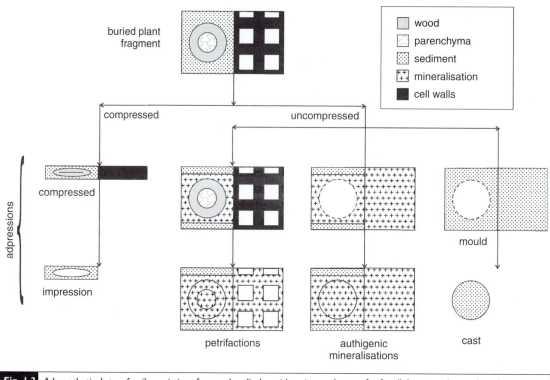

buried plant
fragment

wood
parenchyma
sediment
mineralisation
cell walls

compressed   uncompressed

adpressions

compressed

impression

mould

petrifactions

authigenic
mineralisations

cast

Fig. 1.3 A hypothetical stem fossil consisting of a woody cylinder with an internal core of soft cellular tissue (parenchyma) preserved in different ways. Each mode of preservation is represented by a rectangle, the left-hand side of which represents the stem in transverse section in the sediment, showing the distortion in shape that it has undergone during fossilisation. The right-hand side of the rectangle is a schematic representation of cells in close-up to show the level of anatomical detail that is preserved.

tissue itself is converted to a thin layer composed mainly of carbon and is called a phytoleim. If the phytoleim is still preserved, the fossil is known as a compression, but if the layer is lost either though geological changes (e.g. additional compression and/or heat) or weathering after the fossil has become exposed, it is known as an impression. In many cases, a particular assemblage may include both fossils that retain the phytoleim (i.e. compressions) and others where it is lost (i.e. impressions). For instance, when a specimen is broken open, one half will often carry the phytoleim, while the other half (the counterpart) will just show an impression of the plant fragment. Where the matrix is soft, the phytoleim can often become detached from the rock, leaving parts of the specimen as a compression and other parts as an impression. In such cases, the general term adpression is used.

Fig. 1.4 Different types of preservation in plant fossils. a. Adpression. Scale bar = 10 mm. This specimen of *Paripteris gigantea* (Sternberg) Gothan has some pinnules with the carbonised plant tissue still intact (i.e. compressions), whilst others have lost the tissue (i.e. impressions). Specimen from the Faisceau de Meunière (Middle Pennsylvanian Series) in the Dechy Mine, Douai, northern France (Laboratory for Palaeobotany, University of Lille, France, Specimen 947). Photo by J.-P. Laveine. b. Silicified cast of fossil wood. Scale bar = 20 mm. Its woody texture, including some knots, is clearly visible but without evidence of the structure of the wood, it can be difficult to be sure of the type of tree that it originated from. Specimen from Lower Cretaceous strata near Sevenoaks, Kent, England (NMW Specimen 50.140G.1). Photo by NMW Photography Department. c. Authigenic mineralisation in a siderite nodule of a *Lepidostrobus cone*, from an unknown locality, probably of Middle Pennsylvanian age. Scale bar = 30 mm (NMW Specimen 58.464.G440). Photo by C. J. Cleal. d. Transverse section through a petrified stem of a small Mississippian lycophyte *Oxroadia gracilis* Alvin showing detailed cell anatomy of the stele and cortex. Scale bar = 0.1 mm (Bateman Collection, Specimen OBD(2.15)038bT/2). Photo by R. Bateman.

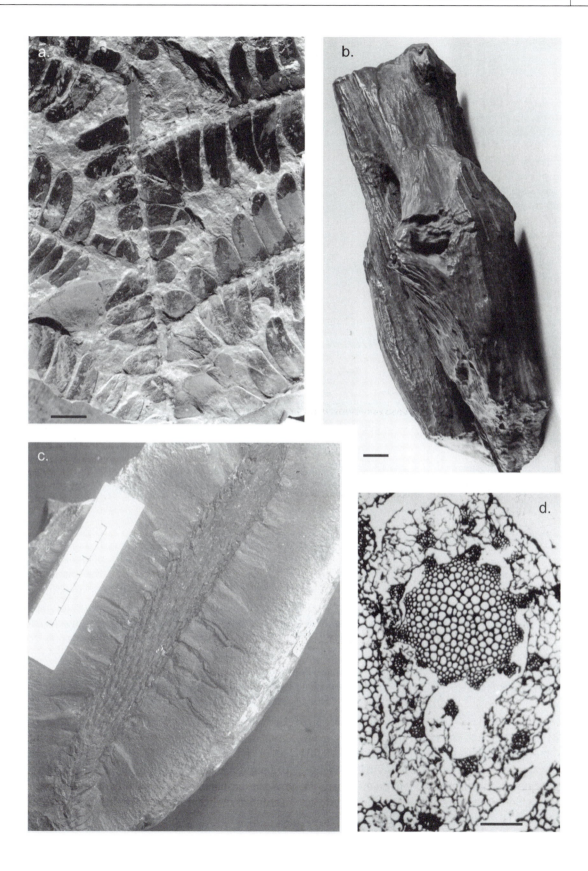

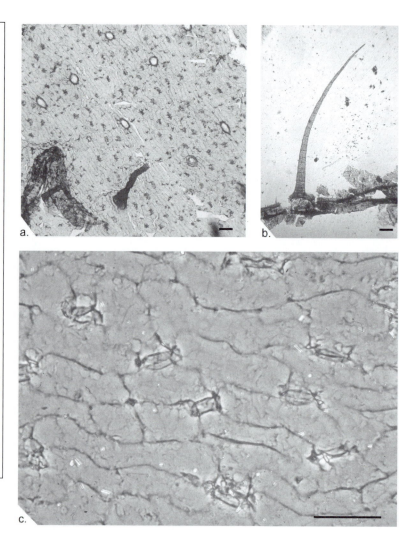

Fig. 1.5 Examples of cuticles prepared from Carboniferous pteridosperm fronds. a. Lower cuticle from *Neuropteris flexuosa* Sternberg frond, stained with safranin, showing hair bases, an attached hair and stomata. Scale bar = 200 μm. Lloyd Cove Seam (Middle Pennsylvanian Series), Brogan's Pit, near Pt Aconi, Cape Breton, Canada (E. L. Zodrow Collection, Specimen 981GF-353). b. Unstained epidermal hair from rachis of *Odontopteris barroisii* Bertrand. Scale bar = 50 μm. Heiligenwald Formation (Middle Pennsylvanian Series), St Barbara Colliery, Saarland Coalfield, Germany (Saarbrücken Mining School Collections, Specimen C/4054). c. Lower (abaxial) cuticle from frond of *Neuropteris ovata* Hoffmann showing stomata. Scale bar = 50 μm. Photographed using phase contrast. Kallenberg Seam, Luisenthal Formation (Middle Pennsylvanian Series), Itzenplitz Colliery, Saarland Coalfield, Germany (Saarbrücken Mining School Collections, Specimen C/3638). Photos by C. J. Cleal.

This process of compression destroys most evidence of cellular-structure. The only notable exception is the cuticle (the outer 'skin' of the plant, Fig. 1.5), which generally survives because of its make-up of relatively non-biodegradable aliphatic polymers. Where the fossil has been subjected to too much heat and/or compression even the cuticle will be destroyed. The cuticle itself does not consist of cells, but often bears the impression of the outer layer of cells of the plant (the epidermis) that it covered. The epidermis includes many structures that are important for understanding the systematic position and ecology of the plants (e.g. epidermal hairs and the stomata or 'breathing pores'), and so the study of cuticles has become an important facet of palaeobotany.

Under certain circumstances the plant fragment may be preserved with little or no compression having taken place. One way this can happen is if the sediment around the plant fragment hardens quickly before significant compression has occurred. The plant tissue itself will subsequently rot away or become reduced to just the carbon residue of the tissue, but there will still be the cavity remaining within the rock that will represent the original shape of the plant fragment. Fossils such as this, known as moulds, are most typically formed around robust parts of plants such as wood and seeds. In some cases, the cavity is subsequently filled up with fine sediment or mineral growth, resulting in a cast of the plant fragment (Fig. 1.4b).

Another type of cast occurs where stems have a central core that is either hollow or made of soft, easily decayed cellular tissue. Sediment may fill the central cavity before the rest of the tissue has decayed producing a sediment-cast of the internal cavity, normally referred to as a pith cast. Sphenophytes and cordaites are particularly prone to this mode of preservation. Sometimes stems are infilled when the remains of the plants are still in the position of growth, which can give an idea of the original density of the plants. Other stems may have been transported away by water and these generally collapse a little before becoming infilled with sediments (see Chapters 6 and 8).

Moulds and casts tend not to preserve fine details of the plant fragments, but there can be exceptions. If the chemistry is right, a piece of plant falling into water and sediment can act as a nucleus for mineral precipitation, and a nodule can form around it; siderite (iron carbonate) is the most common mineral to form such nodules. If this mineralisation happens quickly, the resulting nodule can form a mould around the plant fragment, preserving fine detail, even sometimes including cellular detail of the epidermis and some sub-epidermal layers. Such fossils are known as authigenic mineralisations (Fig. 1.4c). One of the most notable fossil floras preserved in this way is the Pennsylvanian-age Mazon Creek flora of Illinois, USA, which has been the subject of a number of important studies, the most recent being by Wittry (2006).

An analogous mode of preservation is where a plant fragment or small animal is trapped in the resin produced by various types of tree. After the volatile oils have evaporated the resin hardens and is known as amber. Only very small plant fragments are normally preserved in amber, but extremely fine preservation occurs as they are entrapped very quickly; such fossils have proved particularly useful for the study of fossil flowers (Fig. 10.17a).

The best-preserved plant fossils are where little compression occurred and even the cellular detail of the tissue is preserved; these are known as petrifactions (Fig. 1.4d). They form when fluids containing minerals (e.g. calcite or silica) in solution have percolated through the body of the plant before significant decay has occurred. The cells themselves become impregnated by the mineral, which crystallises to preserve their form. The cell wall is sometimes retained as a thin layer of coal around the mineral replacement of the cell contents (cell lumen) or is itself replaced by mineral. Either way, the detailed anatomy of the plant fragment is revealed when a section is cut through the fossil. Examples of petrifactions occur throughout the fossil record of plants. Notable examples include the Early Devonian Rhynie Chert flora from Scotland that has provided critical evidence for understanding the early evolution of land plants (discussed in Chapter 4); the Carboniferous and Permian coal ball floras from the USA, Europe and China that have allowed us to gain a better understanding of the very earliest tropical rain forests; and the Jurassic Rajmahal Hills flora of India, that has clarified many aspects of Mesozoic plants.

Petrifactions only occur under unusual conditions, such as in habitats associated with volcanic activity or where the plant fragments have been soaked in sea water. They are therefore much rarer than adpressions and casts, and tend to represent vegetation growing in extreme habitats. Consequently, if we only looked at petrifactions, we would get a very biased understanding of past vegetation. Where they do occur, however, petrifactions provide critical anatomical evidence that illuminates parts of the evolutionary history of plants and thus complements the more complete adpression record.

Where are plant fossils found?

Plant remains can be found fossilised in most types of sedimentary rock, in a range of different localities (Fig. 1.6). However, the most abundant and best preserved tend to occur in rocks that were formed in non-marine environments. Rocks formed from sediment deposited in river deltas provide some of the best opportunities to find plant fossils, especially the finer-grained sediment deposited in lakes within the deltas. If the water table remains generally high in the delta sediment, plant fragments will be much

Fig. 1.6 Plant fossil sites. a. Coastal exposure at Cayton Bay, Yorkshire, England. Collecting at this classic Middle Jurassic plant bed in 1970 (left to right: J. Watson, K. Sporne, M. Boulter and M. Mortimer). b. Exposure at Edrom on the banks of the Whiteadder River, Berwickshire, Scotland. This site has yielded a diverse assemblage of Mississippian plant petrifactions, especially of early seed-plants. c. The field at Rhynie in Scotland, that has yielded the classic Early Devonian macroflora discussed in Chapter 4. There are normally no exposures of this plant bed, but there have been occasional excavations to collect further material (e.g. Fig. 2.11). d. Road cutting near Beckley, West Virginia, USA. Bill Gillespie (centre) and Mitch Blake collecting from the Early Pennsylvanian Pocahontas No. 2 Coal. e. Disused clay pit at Czerwionka-Leszczyny, Upper Silesia, Poland, where Middle Pennsylvanian plants can be found. f. Spoil tip from abandoned underground coal mine. Middle Pennsylvanian flora from near Radstock, England, being collected by Deborah Spillards and Chris Cleal. Photos by B. A. Thomas (a, f) and C. J. Cleal (b–e).

Fig. 1.7 Middle Pennsylvanian sequence at Point Aconi, Cape Breton, Canada. This shows a coal seam (the Point Aconi Seam), underlain by a seat earth (fossil soil), and overlain by mudstones that were deposited when water and sediment flooded the area and destroyed the peat-forming vegetation. Photo by B. A. Thomas.

the fossil record such sedimentary changes have resulted in the formation of seams of coal separated by sandstones and mudstones containing plant fossils. The prime examples of this type of sequence can be found in the Pennsylvanian coal-bearing sequences ('Coal Measures') of Europe and eastern North America.

Although the coal/peat consists almost exclusively of plant remains, it is usually very difficult to study them because they are all crushed together in a confused mass, so one can rarely discern the shape of any of the pieces of plant that make up a lump of coal. It is possible to macerate coal using strongly oxidising acids so that some of the plant material can be extracted, but the remains are usually so broken-up that they are difficult to study. Pollen, spores and cuticles are the most common types of fossils studied in this way. The only circumstances where anatomical details of the component plant remains of coal can be studied are when the seam has been impregnated with mineral matter, petrifying some of the plant tissues. Such mineralisation is not widespread but where it does occur it can produce extremely fine preservation of the plant tissue. The best known examples of such preservation are the Pennsylvanian-age coal balls of Europe and North America (examples of such coal ball petrifactions are shown in Figs. 5.11, 7.6c, 8.11e).

If the water table is lower, plant decay is much quicker and peat tends not to build up. Low water table conditions can be recognised by the oxidisation of the sediment, often resulting in red-coloured rocks (red beds). Plant fragments may still become encapsulated in the sediment in such situations, but on the whole they are less common than when the water table has been higher. If they are found in such red-beds, the carbon phytoleim is usually lost and the fossil is preserved as an impression.

Volcanic landscapes tend to be very unstable and their sedimentary deposits are often subject to considerable reworking. On the face of it, this would seem to be an unlikely setting in which to find plant fossils. However, the ground waters in such environments are often mineral-rich and so can petrify plant fragments. An example of this is the Rhynie Chert deposit, formed when an

slower to decay and thus stand a greater chance of being preserved within the sediment. If both sedimentation rates and decomposition of the plant litter are slow, peat can accumulate and over geological time this can result in coal (Fig. 1.7). If sedimentation rates are more rapid, the plant fragments will be buried in sands or mud, and when these deposits are turned into solid rock the plant remains are preserved, usually as adpressions or casts, as described above. Both rapid and slow sedimentation rates can occur in repeated sedimentary successions, such as the cyclothems that result from a rhythmic raising and lowering of sea-levels during glacial-interglacial cycles. A number of times in

Early Devonian peat accumulation was inundated by mineral rich waters, preserving in exquisite detail the internal anatomy of some of the very early land plants (see Chapter 4).

Lagoonal and coastal deposits may contain fragments of plants that had drifted from the coastal vegetation. An example here is the Jurassic Stonesfield 'Slate' flora of Oxfordshire, where plant fossils including conifer and fern fragments are found with the remains of marine animals such as bivalves. Another is the Sheppey flora from Kent, which has Tertiary plant remains associated with sharks' teeth, crabs and other animal remains. The diversity and preservation of other drifted floras is sometimes much poorer although coastal and lagoonal deposits may preserve remains of vegetation that is different from that found in deltaic deposits of the same age. For instance, the Stonesfield flora is quite different from that found in the deltaic deposits in Yorkshire, despite being of similar Middle Jurassic age.

Bias in the fossil record

Although fossils are the only direct record that we have of organisms that lived in the past, they provide a very incomplete picture. This problem has been known about since the nineteenth century and Charles Darwin discussed it extensively in his book *On the Origin of Species*. The fossil record provides a particularly incomplete picture of past vegetation. It is probably at its best when plants were first starting to migrate onto land, in Silurian and Early Devonian times, as the wet habitats being colonised were well suited for fossilisation. As soon as plants adapted to drier habitats, bias in the record becomes significant and only those plants that continued to occupy wetter habitats, such as riverbanks and deltas, and lake shores, are well represented as fossils. Since many of the main evolutionary developments in plants have been adaptations to life in drier regimes (especially reproductive adaptations such as seeds and flowers, see Chapters 8–10), there is a very real problem in using the fossil record directly to develop models that explain the evolutionary history of plants.

There are two main types of evidence that suggest the fossil record is incomplete. Firstly, sedimentary rocks sometimes yield tiny fragments of plants that have travelled a much greater distance than the larger specimens that make up the vast bulk of the plant fossil record, and probably represent the vegetation growing in more elevated and thus drier habitats. Charcoal (the product of fire and technically called fusain), for instance, can travel great distances along rivers and can often preserve surprisingly fine morphological details of plants including some internal anatomy. Spores and pollen grains can also travel vast distances and give us information about the plants growing in drier habitats away from the sedimentary basins. The problem here, however, is that we do not know which groups of plants produced some of the spores and pollen grains. Nevertheless, from both of these sources of information we are fairly certain that in Late Carboniferous times there were forests in upland areas with various seed-plants, including conifers and cycads, long before those plants are represented in the main fossil record.

There are also indirect means of judging the incompleteness of the plant fossil record. One example is known as 'the molecular clock'. By determining the rate at which genes randomly mutate, it is possible to use the number of genetic differences between two organisms to estimate how long ago they shared a common ancestor. Such evidence suggests that the flowering plants diverged from the other seed-bearing plants as long ago as Triassic times, if not earlier. However, this is in marked contrast to the evidence of the fossil record, in which the oldest unequivocal occurrence of flowering plants is in rocks of early Cretaceous age. So, either the molecular clock has been 'running fast', or there were flowering plants growing in upland areas for millions of years before they appeared in the lowland habitats that are sampled by the fossil record.

Even when trying to interpret the vegetation that gave rise to a particular fossil assemblage, bias in its composition becomes a real issue. The plants will have been subjected to varying degrees of damage due to fungal or animal attack

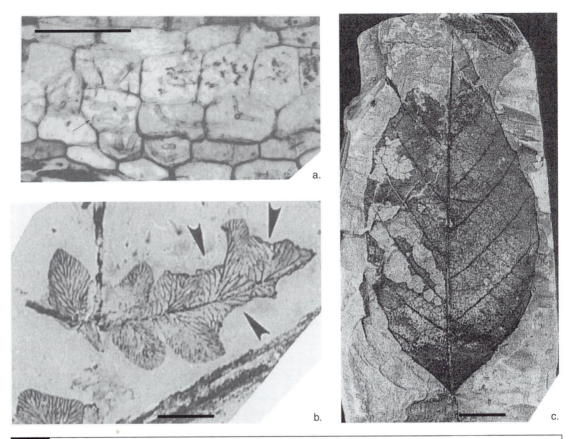

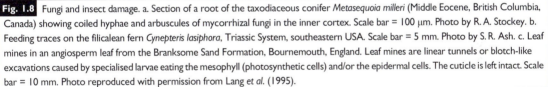

Fig. 1.8 Fungi and insect damage. a. Section of a root of the taxodiaceous conifer *Metasequoia milleri* (Middle Eocene, British Columbia, Canada) showing coiled hyphae and arbuscules of mycorrhizal fungi in the inner cortex. Scale bar = 100 μm. Photo by R. A. Stockey. b. Feeding traces on the filicalean fern *Cynepteris lasiphora*, Triassic System, southeastern USA. Scale bar = 5 mm. Photo by S. R. Ash. c. Leaf mines in an angiosperm leaf from the Branksome Sand Formation, Bournemouth, England. Leaf mines are linear tunnels or blotch-like excavations caused by specialised larvae eating the mesophyll (photosynthetic cells) and/or the epidermal cells. The cuticle is left intact. Scale bar = 10 mm. Photo reproduced with permission from Lang et al. (1995).

(Fig. 1.8) that may affect different groups of plants in different ways. Delicate plants such as mosses and liverworts are for this reason rarely represented in the fossil record, and consequently our understanding of the evolutionary history of these groups is much poorer than that of more robust plants such as ferns.

Most assemblages are allochthonous, often derived from two or more original habitats, and these may not be representative of the overall vegetation of the area. For instance, in the Pennsylvanian Coal Measures of Europe and North America, most fossils are found in the mudstones and siltstones, and represent a mixture of the vegetation that grew along the banks of these palaeotropical rivers. However, the vast bulk of the forests consisted of giant lycophytes ('club mosses') growing in the swamp areas behind the raised levees and these found their way into the sediment nowhere near as often as the riparian plants. Only by studying the hypautochthonous peat (now converted into coal) have we been able to get a more representative picture of what the Coal Forests as a whole looked like (Fig. 1.9).

Autochthonous and hypautochthonous assemblages may sometimes give a more representative picture of the vegetation that grew in an area, although even here bias exists. For instance, where an area of forest has been killed by being inundated by flood water and sediment leaving the stumps of the larger trees in place (resulting in what we call a fossil forest) the

Fig. 1.9 Reconstructions of the Pennsylvanian palaeotropical Coal Forests. a. Diverse vegetation growing on the river bank, from where the bulk of the plant fragments found preserved as adpressions were derived. b. A more representative view of the Coal Forests, showing that most of the vegetation consists of arborescent lycophytes, and the riverbank vegetation is only a minor component. Drawings by A. Townsend.

understory and climbing plants are usually washed away.

Most plant fossil assemblages should not, therefore, be confused with a vegetation assemblage (flora) as a modern-day botanist would see it. To express this difference between a plant fossil assemblage and a living flora, palaeobotanists have coined various terms over the years, none of which has received universal acceptance. In this book we use the term macroflora for an assemblage of plant macrofossils (i.e. fossils that can be seen with the naked eye, in contrast to microfossils that have to be examined under the microscope) found in a particular locality and in a particular stratigraphical unit, but we do not define localities or stratigraphical units. Similarly, an assemblage of pollen and spores extracted from a particular stratigraphical unit at a particular unit is described as a palynoflora.

Why do we study plant fossils?

There is evidence that man had come into contact with the fossilised remains of plants even in prehistoric times; for example, a piece of fossilised wood was found in a Scottish neolithic hearth, which had presumably been placed there by an unsuspecting prehistoric cook! The serious study of plant fossils, however, started in the eighteenth century, and developed as a major discipline in the nineteenth century (for a more detailed discussion of this subject, see Chapter 2).

The fossil record, even though incomplete and biased, is the only direct means of finding out what plants grew in the past. Most other sources are based around extrapolations into the past from present-day data, which require many assumptions in the analysis. With the fossil record, we can see the actual remains of extinct plants that lived in the past and, even with its limitations, it still has much to tell us of the evolutionary history of plants.

There is other information that the record can provide. Biogeographical studies can help in palaeogeographical reconstructions. The apparent anomalies in the distribution of Palaeozoic plant fossils relative to the distribution of the continents today was one of the arguments used by Wegener in developing his continental drift model in the 1920s. Wegener's hypothesis formed the basis for the plate tectonics model now used to explain many large-scale processes in earth sciences. This topic will be dealt with in more detail in Chapter 11.

Changes in the distribution of floras through time can also be an extremely valuable proxy for past climate change, as can changes in leaf shape (leaf physiognomy), especially among the angiosperms. Variations in stomatal density have been used to estimate both long- and short-term fluctuations in atmospheric CO_2. Even the presence of charcoalified plant fossils is now thought to provide an important constraint on estimates of atmospheric O_2; there cannot have been wildfire (which is thought to generate most charcoal) if atmospheric O_2 was below a certain level.

Plant fossils can be used to estimate the relative ages of the rocks in which they are found by comparing the patterns of appearances and disappearances through the stratigraphical successions – a science known as biostratigraphy.

Animal fossils such as ammonites and corals are generally used for such work in marine strata. Animal groups have also been used in non-marine deposits (good examples are the non-marine bivalves in the Pennsylvanian successions of western Europe). Generally, however, plant fossils are of more use in non-marine rocks, especially those of the Late Palaeozoic age, where there seems to have been a very rapid turnover of species. The distribution of plant and spore fossils in non-marine and near-shore marine deposits has been particularly important in helping to correlate the two environments.

There can also be a simple economic imperative to study plant fossils. Their biostratigraphical value has proved useful in exploration for natural resources such as oil. Plant remains are also the basis of one of the world's most important energy resources – coal. Most coals are the remains of peat generated in swamps and forests, and maximising their exploitation can depend on understanding how that peat was formed. This in turn depends on understanding the vegetational dynamics of the original forests, which can only be determined by the study of the plant fossils.

Plant fossils are, therefore, an important tool for the botanist trying to understand the evolution of plant-life, the geologist wanting a means of correlating strata and establishing past continental positions, the climatologist who wants to know about past climates and atmospheres, and the mining engineer who needs help in the exploitation of coal reserves. One reason for the continuing fascination of the subject is that it relates directly to so many different fields.

Recommended reading

Banks (1964), Cleal (1991), Delevoryas (1966), Gordon (1935), Jones & Rowe (1999), Meyen (1987), Rex (1993), Rex & Chaloner (1983), Schopf (1975), Stace (1989), Stearn (1992), Stewart & Rothwell (1993), Taylor & Taylor (1993), Thomas & Spicer (1987), Walton (1936).

Chapter 2

Highlights of palaeobotanical study

Plant fossils have been studied for over 300 years and today there are several hundred active palaeobotanical researchers around the world. This chapter is a somewhat subjective summary of how the subject has developed, focussing on a number of themes that we believe have been significant. Some more complete historical reviews are listed at the end of this chapter.

The beginnings of palaeobotany

Although there has been an awareness of the existence of plant fossils since pre-historic times, the earliest detailed, scientific descriptions date back to the late seventeenth and early eighteenth centuries, notably by Edward Llwyd (1660–1709), Martin Lister (1638–1711), John Ray (1627–1701), John Woodward (1665–1722) and Johann Scheuchzer (1672–1733). Through their publications they stimulated thought on whether these fossils were really the remains of once living organisms or just mineral accidents. The general reasoning came down on the side of once living organisms, although the religious belief at the time constrained ideas and committed people to believing that fossils were formed as a result of the biblical flood. Nevertheless, there was a general acceptance that the plant fossils could be recognised as species, even though it was thought that they could all be related to living species.

Progress in palaeobotany was slow during the rest of the eighteenth century with only a few publications of any substance. It was still the time of the amateur naturalist, and such men as James Parkinson (1755–1824) and William Martin (1767–1810) were adding to the general knowledge of the range of plant fossils that could be found. James Parkinson was in fact a physician and was the first person to recognise the disease that he called shaking palsy, but which is now known as Parkinson's disease. However, he was also deeply interested in palaeontology and his *Organic Remains of a Former World* (published between 1804 and 1811 in three volumes) was the most important study up until then on British fossils. Parkinson was most notable for arguing that coal was derived from plants and that fossils were important for stratigraphical correlation, although his understanding remained hindered by his belief in the biblical flood. At about this time, there was increasing knowledge of the natural world as explorers came back from around the world with both dried and living plants for the new museums and botanic gardens. A number of these explorers noted the similarity between some of the plants that they were finding in the tropics and those found fossilised in the older, Palaeozoic rocks, and attempted to classify the fossils in this context. Perhaps the most notable of those botanists was Karl von Martius (1794–1868) of Regensburg, who compared Carboniferous plant fossils with the modern-day Brazilian flora. Although this approach ultimately proved misguided, it did open people's eyes to the idea that the early (Palaeozoic) vegetation of Europe was quite different from the vegetation growing there today.

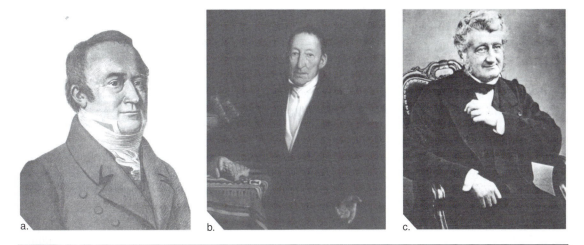

Fig. 2.1 Three early nineteenth century pioneers of scientific palaeobotany. a. Ernst von Schlotheim (1764–1821). b. Kaspar Maria von Sternberg (1761–1837). Reproduced by permission from J. Kvaček. c. Adolphe Brongniart (1801–1876).

Palaeobotany eventually became more 'professional' mainly through the work of three Europeans. The earliest was Ernst von Schlotheim (1764–1821: Fig. 2.1a), who wrote many palaeobotanical works with the most important being *Die Petrefactenkunde* published in 1820, just a year before his death. Schlotheim had collected throughout Germany and France, and clearly understood the stratigraphical significance of his fossils. Although his work was ground-breaking in many ways, it had one major drawback. The taxonomic names that he used to name his fossils tended to be rather poorly defined, and were often more like short descriptive terms rather than formal names in a modern taxonomic sense. This has made it very difficult to incorporate Schlotheim's nomenclature into palaeobotanical taxonomy as used today. To avoid the potential confusion that Schlotheim's names could introduce into the subject, the *International Code of Botanical Nomenclature* now specifically excludes his names as having any validity.

The second was Kaspar Maria Graf von Sternberg (1761–1837; Fig. 2.1b) whose publication *Versuch einer geognostich-botanischen Darstellung der Flora der Vorwelt* published in eight parts between 1820 and 1832 is now accepted as the first to have nomenclatural validity. Sternberg was a wealthy aristocrat, but was no mere dilettante – he was a recognised authority on the botany of the Saxifragaceae, and had built up an extensive collection of plant fossils of all ages. His collection ultimately formed part of the National Museum in Prague, which he was instrumental in founding. Sternberg largely based his sumptuously illustrated book on plant fossils in his collection and in it he developed a number of innovative ideas, notably that there were three distinct periods of vegetation: the periods of coal plants, of cycads and of flowering plants – effectively the Palaeozoic, Mesozoic and Cenozoic as accepted today. Although these 'vegetation types' are now recognised as over-simplified concepts, his introduction of the idea that there had been progressive change in vegetation through stratigraphical succession was a fundamental advance in the study of plant fossils.

The third person was Adolphe Brongniart (1801–1876; Fig. 2.1c) who, like Sternberg, was interested in living plants. Their botanical knowledge influenced both their palaeobotanical studies, although it is perhaps most obvious in Brongniart's greatest work, the *Histoire des Végétaux Fossiles* (published in 15 parts between 1828 and 1838), in which the fossils are classified with a system similar to that used for living plants. Unfortunately the work was never completed (the last published part ends abruptly, in mid sentence) and did not deal with plants that he regarded as seed-bearing (notably conifers, cycads and angiosperms). However, within those groups that he did cover, it was the most detailed, systematic and wide-ranging account that had been published, based on what was then probably the most extensive collection of plant fossils anywhere in the world – now in the *Museum d'Histoire Naturelle* in Paris.

A point to consider here is the length of time over which the parts of these major works were published. As each part came out it reflected both an increased knowledge from new discoveries and the interplay of ideas from other publications. Two important British publications also appeared at this time. Edmund Artis's *Antediluvian Phytology* appeared in 1825 and included comparisons with fossils figured by Schlotheim, Sternberg and Brongniart, remarks on the systems used by these authors, and comments received from the renowned geologist William Buckland, who will be discussed later. Artis even named one of his fossils *Sternbergia* after Sternberg and Sternberg reciprocated with *Artisia*. Unfortunately, Artis was of relatively humble origins being initially a confectioner and then butler to the Whig politician, Earl Fitzwilliam. Because of this he was not accepted by either the British geological establishment or the small international palaeobotanical community. Perhaps not surprisingly, he quickly gave-up the study of plant fossils in preference for his other passion – Roman archaeology – for which he received wider recognition and is better known today. Nevertheless, Artis's *Antediluvian Phytology* stands out as one of the finest illustrated books on plant fossils of its time (Fig. 2.2).

The *Fossil Flora of Great Britain* by John Lindley (1799–1865) and William Hutton (1797–1860) was published in a number of parts between 1831 and 1837. This work contained 230 plates and descriptions of plant fossils ranging in age from Carboniferous to Cenozoic. Lindley was a more accepted figure than Artis (he was the Professor of Botany at the University of London) and so his work with Hutton received greater recognition from the scientific establishment of the time, even though it was not so well illustrated. However, Lindley clearly had increasing problems relating the plant fossils to his knowledge of living plants and seems to have progressively lost interest in the project; it is notable that his discussions on the fossils became briefer and less analytical in later parts of the book. Ultimately, he gave up the project, to return to his studies on living plants. It was 40 years before George Alexander Louis Lebour (1847–1918), then the Professor of Geology at Newcastle upon Tyne, published the rest of the Lindley and Hutton plates.

Heinrich Göppert (1800–1884) was the next important continental palaeobotanist, who started publishing on plant fossils in 1835. His important *Systema Filicum Fossilium* was published in 1836, which included a 76 page history of palaeobotanical endeavours so far. Göppert published widely and in his thesis (1838) figured pollen that he had recovered from a fossil alder catkin. Soon after this, in 1844, Wilhelm Schimper (1808–1880) of Strasbourg published on the Triassic plants of the Vosges. Then, after working on living bryophytes for many years, Schimper returned to palaeobotany with his monumental *Traité Paleontologie Végétale* published in three volumes together with an atlas of plates between 1869 and 1874. It was conceived as a complete manual of systematic palaeobotany with what a contemporary described as, a highly scientific and rational classification system.

Palaeobotany was on a much surer footing in the latter part of the nineteenth century, but many constraints on thinking still prevailed at the time. Religion remained a very powerful influence, and many people thought Darwin to be a heretic when in 1859 he published his ideas about evolution in *On the Origin of Species*. The other major constraint was in the acceptance of the stability of the world as it is today. Until Wegener's continental drift theory (which will be briefly discussed later in this chapter) was accepted, the only way to explain some of the apparently anomalous patterns of distribution of fossil floras had been through the ideas of land bridges between continents, for which there was no physical evidence.

The importance of coal

The nineteenth century saw momentous changes with the industrial revolution sweeping

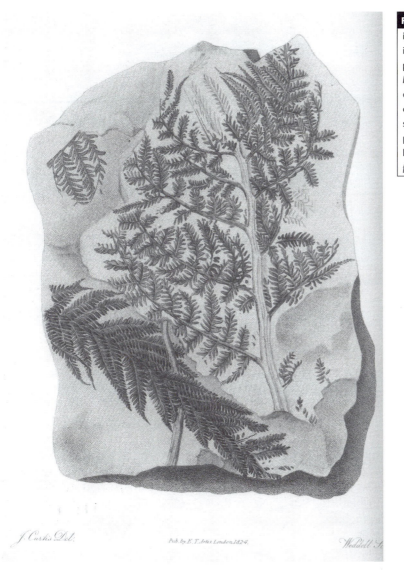

Fig. 2.2 One of the many fine illustrations of plant fossils included in Artis's classic 1825 palaeobotanical book *Antediluvian Phytology*. Before the development of photography, plant fossils could only be figured using engravings, such as this, or lithographs. This particular example is the holotype of *Senftenbergia plumosa* (Artis) Stur.

through Europe. The new industries were powered by coal, so it is not surprising that great effort was put into discovering new coalfields and working all the profitable seams within them. The plant fossils found in the roof shales of many of the coal seams quickly passed from being mere curiosities to providing a possible means of comparing the age of different coal fields and the stratigraphical positions of their various coal seams. Palaeobotany therefore attracted many able people who were often associated with coal mining, surveying or transport.

Eventually such age comparisons were worked out not only within Europe and later North America, but also between Europe and North America.

The Frenchman Francois-Cyrille Grand'Eury (1838–1917) started working life as a mining engineer, but later became the Professor of Trigonometry at St-Étienne. One of his greatest contributions was in assembling whole plants from their fragmented fossilised remains. He did so through careful observation and judging the evidence cautiously before committing

a. b.

Fig. 2.3 a. Robert Kidston (1853–1924: left) and Thomas Gwynne-Vaughan (1871–1915) at work in Kidston's laboratory in his house at Stirling, Scotland. b. From left to right: John Walton (1895–1971), Frederick E. Weiss (1865–1953) and William H. Lang (1874–1960), photographed in 1930. Photo provided by J. Watson.

himself. His reconstruction of the cordaite tree is still used in publications today. Grand'Eury was, however, convinced of the necessity for using distinct generic names for the different organs even if he was convinced that in this instance they were parts of the same original plant.

Another Frenchman René Zeiller (1847–1915) was one of the first people to use plant fossils seriously in Carboniferous and Permian stratigraphy, publishing his three large memoirs in 1884, 1888 and 1906. Zeiller had a great interest in Carboniferous ferns and was also the first to describe the male fructifications of pteridosperms.

Robert Kidston (1853–1924; Fig. 2.3a) was a contemporary of Zeiller but, unlike him, Kidston did not have to work for a living. At the age of 26, Kidston became financially secure and devoted the rest of his life to palaeobotany. In 1880 he was appointed to the position of Honorary Palaeozoic Palaeobotanist to the British Geological Survey, on the retirement of the botanist, Sir Joseph D. Hooker (1817–1911). This gave him an unrivalled opportunity to receive plant fossils from all over the country from the Geological Survey. Kidston also received specimens from the professional collector William Hemingway of Barnsley in the Yorkshire coalfield. His main interest was

in the Coal Measures and he published many accounts of the morphology of plant fossils from British coalfields, including detailed studies on the sporangia of the various ferns found there. When his interest turned to the anatomy of fossil plants, however, he needed a co-worker. This was Thomas Gwynne-Vaughan (1871–1915; Fig. 2.3a) who was an assistant to Frederick Orpen Bower (1855–1948) the Professor of the Botany Department at the University of Edinburgh. Kidston and Gwynne-Vaughan became friends, and worked together in Edinburgh, and in Stirling, where Kidston had set up a laboratory in his house. Gwynne-Vaughan moved several times, becoming Head of Department at Birkbeck College in London, the inaugural Professor of Botany at Queen's University in Belfast, and in 1914 Professor of Botany at Reading. During this time, Kidston and Gwynne-Vaughan published their classic series of papers on the fossil Osmundaceae and it is clear that they would have gone on to study the Rhynie Chert together (see below) but Gwynne-Vaughan died of tuberculosis before the work could start.

In the early 1920s the Geological Survey allocated funds to publish an illustrative catalogue on Carboniferous plants and commissioned

Fig. 2.4 E. A. Newell Arber (1870–1918). Drawing by his wife Agnes Arber, who became a highly recognised botanist in her own right. With permission from M. Arber.

intended, they would have contributed greatly to the Coal Measures stratigraphical work that was being done at the time. However, in the intervening years, Carboniferous palaeobotany had undergone considerable advances, especially through the work of continental palaeobotanists, and the 'Crookall' monographs did not take this into account.

Working at about the same time as Kidston, was E. A. Newell Arber (1870–1918; Fig. 2.4) who was the Demonstrator in Palaeobotany at the Sedgwick Museum, Cambridge. Arber investigated a range of fossil floras, but his most significant contributions were accounts of Pennsylvanian floras from several British coalfields. He concentrated on adpressions, rather than on the type of petrifactions that were attracting increasing attention at that time, as he believed that they were a much more important source of information on past floras. Like Kidston, Arber was a great believer in the importance of fieldwork for expanding our knowledge of extinct floras. His other main interest was modern alpine vegetation, which came about through spending time convalescing in Switzerland as a teenager. Arber suffered from the effects of an early attack of pneumonia all his life and he died at the relatively young age of 48.

People like Emily Dix (1904–1973; Fig. 2.5) and later Leslie Moore (1912–2003) actively worked on the stratigraphy of South Wales. Dix was trained by the eminent British stratigrapher, Arthur Trueman, who had developed the use of biozones for interpreting the stratigraphical distribution, initially of Mesozoic ammonites, and later of Carboniferous bivalves. In her pioneering 1934 paper, Dix was the first to use this biozonal approach with the plant fossil record, originally in the South Wales Coalfield, and later in other British coalfields. Unfortunately Dix suffered a mental breakdown in the late summer of 1945 from which she never recovered. Before this, however, she had managed to establish a close working relationship and friendship with the leading authorities of the day, notably Paul Bertrand (1879–1944) of France, William Culp Darrah (1909–1989) of the USA, Walther Gothan (1879–1954) of Germany, Wilhelmus Jongmans (1878–1957) of The Netherlands, and Armand Renier (1876–1951) of

Kidston to do the work. This led to his immensely important monograph on Coal Measures plants, published in six parts between 1923 and 1924, which figured and described the very best of his superb collection (now in the collections of the British Geological Survey in Keyworth, Nottingham). Unfortunately, Kidston died in 1924 while visiting colliery agent and amateur palaeobotanist, David Davies (1871–1931) in South Wales who was using his own enormous collection (now in the collections of NMW) for stratigraphical and ecological interpretations of the South Wales Coalfield. The remaining six parts of Kidston's monograph were completed by Robert Crookall (1890–1981) and published between 1955 and 1970 (some even after Crookall himself had retired), but they appeared too long after the core of the text had been written by Kidston and were thus seriously out of date. If they had been published in the 1920s, as

Fig. 2.5 Photograph of palaeobotanists at the 1930 International Botanical Congress. Included here are many of the leading Palaeozoic palaeobotanists from the years between the two world wars. Photo provided by J. Watson.

Belgium (many of these are shown in Fig. 2.5). Through their own work, their co-operation in exchanging plant fossils and exchanging ideas, with the occasional disagreement, they extended their knowledge of Carboniferous plants and stratigraphy beyond their own country boundaries, and devised floral sequences to determine the comparative ages of Coal Measures strata.

This work should be set against the changing situation in Europe. The 1914–1918 war had finished and Europe was at peace again, although conditions were, of course, different in Russia because of the communist revolution in 1917. International co-operation was made easier with more modern transport and people were travelling greater distances. In 1927 the first of the congresses on Carboniferous stratigraphy was held in Heerlen, The Netherlands, and these important congresses have continued ever since with the sixteenth held in Nanjing, China in 2007. Stratigraphical research continues, of course, and there is much still to be understood. In North America, Charles Read (1907–1979) and Sergius Mamay (1920–2008) developed an Upper Palaeozoic floral biostratigraphy that is still used by some palaeobotanists in the USA. In Europe, Robert Wagner (a student of Jongmans) recognised in the late 1970s that there was a significant stratigraphical gap between the top of the Westphalian Stage and the base of the Stephanian Stage as traditionally recognised. To fill this gap, he proposed a new Cantabrian Stage (now Substage). In 1984, Wagner devised a new set of biozones for Carboniferous floras, which attempted to integrate the schemes developed in Europe by Dix, and Read and Mamay in the USA, with his own observations, principally in Spain.

Anatomical studies

Palaeobotanical studies have not been entirely focussed on stratigraphy, for many palaeobotanists preferred to study their fossils botanically.

New techniques were developed to help this aspect of plant fossil investigations. John Walton (1895–1971; Figs. 2.3b, 2.5) devised a method for 'turning fossils over' by sticking them facedown onto glass slides and dissolving the rock away with acid (a very similar technique had, in fact, been tried by Artis in the early nineteenth century, although this aspect of his work has tended to be overlooked). Walton's technique can reveal reproductive organs on the underside of leaves, which are generally concealed in the rock. This is because when the rock is split the fissure plane usually runs along the smoother surface of the leaf (usually the upper surface).

Cuticles also held promise for helping to improve the understanding of Carboniferous plants. Although their value had been clearly demonstrated in Mesozoic floras during the first half of the twentieth century (see later in this chapter) their use for Carboniferous palaeobotany only really started to develop in the 1960s, mainly through the efforts of the German palaeobotanist Manfred Barthel with pteridosperms, and Barry Thomas with lycophytes. Today, however, cuticles have become accepted by most palaeobotanists working on Carboniferous compression floras as an integral part of their study.

Fig. 2.6 William Crawford Williamson (1816–1895). From a reprint of Williamson's 1896 autobiography *Reminiscences of a Yorkshire Naturalist*.

Coal balls

The emphasis of most of the work discussed so far has been on adpression material, which of course gives the collector an instant picture of what the plant (or at least, plant parts) looked like. However, to determine the anatomy of a fossil plant, which is often necessary to establish its systematic position, it is essential to investigate plant petrifactions.

The first studies on anatomically preserved plant fossils were undertaken by two British geologists, William Nicol (1770–1851) and Henry Witham (1779–1844), both of whom published accounts of 'petrified wood' in the early 1830s. They used the pioneering technique of grinding thin, transparent sections that were mounted on glass slides for microscopic study. But it was the discovery in the mid-nineteenth century of calcareous nodules (coal balls) in

some coal seams with plant remains preserved in exquisite anatomical detail that stimulated the study of fossil plant anatomy. The pioneer in this work was Edward Binney (1812–1881), who published the first major paper on the subject in 1885 with the great botanist of the time Joseph Hooker (1871–1911). They showed that all kinds of plant tissues could be preserved in this way and how sectioning could reveal important anatomical detail, including details of reproductive structures. This opened up a whole new field for anatomical study that has yielded so many immensely important discoveries of botanical and evolutionary significance.

Coal ball study was later taken up in an enthusiastic way by William Crawford Williamson (1816–1895; Fig. 2.6). As a youth in Scarborough,

Yorkshire, Williamson had dabbled in archaeology, ornithology and palaeontology, and contributed drawings of Jurassic plant fossils to Lindley and Hutton's *Fossil Flora*, referred to earlier. He went on to have a varied and successful career in medicine, which he practiced all his life, and pioneered work on the structure and development of teeth and bone, which gained him a Fellowship of the Royal Society of London in 1854. In 1851, he became the first Professor of Natural History at what was then Owen's College in Manchester, and in 1858 he published an important monograph on living foraminifera based on the microscopic study of sea-bed samples from all over the world. Ultimately, however, Williamson became best known for his 19 important palaeobotanical memoirs on coal ball petrifactions, perhaps the most memorable being his study on the lepidodendroid rooting base *Stigmaria* (see Chapter 5, especially Fig. 5.20b). In this work, he was able to show that the trees of the Carboniferous were giant spore-producing, rather than seed-bearing, plants. The results of these extensive anatomical studies earned Williamson a Royal Medal from the Royal Society in 1874. Williamson collaborated later in life with Dukinfield Henry Scott (1854–1934), who went on to become one of the leading palaeobotanists of his time. Williamson's (1896) autobiography makes interesting reading, describing his life, the times and the people with whom he worked.

Coal ball studies were continued in Britain after Williamson's retirement by D. H. Scott, and by Williamson's successor at Manchester, Frederick E. Weiss (1865–1953). The latter also encouraged a number of younger palaeobotanists to investigate these petrifactions, including William Lang and John Walton (Fig. 2.3b). However, these younger palaeobotanists soon started to drift towards other floras and interests (to be discussed later) and coal ball studies in Britain underwent a marked decline in the 1920s, never properly to recover.

Coal ball studies continued in continental Europe during the first half of the twentieth century, perhaps most notably those by the Belgian Suzanne Leclercq (1901–1994). However, it was in North America that their study underwent the next major developments. Coal balls were first recorded from the USA in the late 1890s although these were pyritic and could only be studied using reflected light. But in the 1920s, carbonate coal balls were discovered in the Illinois coalfield that could be studied in the same way as those that had been found earlier in Europe. Adolphe Noé (1873–1939) was the first to undertake palaeobotanical investigations in the early 1920s aided by his students, most notably J. Hobart Hoskins (1896–1957). Many of these American coal balls are stratigraphically younger than those found in Europe, and so often contain different species. However, their study initially was subject to the same limitations as the European's had experienced: although ground thin sections could show fine cellular detail, without being able to cut repeated serial sections through the fossil it was very difficult to determine the three-dimensional shape of the structures being investigated. John Walton re-enters the story here by devising, in 1923, the peel technique for making serial sections through the fossil. This was a great leap forward in coal ball study because it permitted the details of the smallest of plant organs to be examined with the minimum of waste. Walton's original technique used cellulose compounds that took a long time to dry and so making repeated sections was a time-consuming business. This all changed in 1956, however, when William (Bill) Lacey (1923–1995) and his co-workers introduced the idea of using acetate sheets to make sections very quickly (see Chapter 3 for a discussion on these techniques).

It was in America that the value of Walton and Lacey's technical innovations was properly realised. From the 1950s there was an explosion of studies in Carboniferous coal balls, most notably by Henry Andrews (1910–2002), Aureal Cross, William Culp Darrah (mentioned earlier because of his adpression work), James Schopf (1911–1978), Wilson Stewart (1917–2004), and more recently by Tom Phillips, Tom Taylor and Gar Rothwell, and their various colleagues and students. There is also a tremendous resource of coal balls from the Donetz Basin in the Ukraine, although these have not been studied in the same detail. In recent years, new discoveries of Permian coal balls in China are now stimulating the anatomical study of these somewhat younger plant fossils, and have the potential to improve our understanding of the evolutionary changes that were taking place.

The peel technique is not limited to coal balls. Albert Long (1915-1999) also used it to great effect on seeds and stems found in Mississippian petrifactions. His early work was on coal balls in William Lang's laboratory in Manchester but, after a period of amateur entomology while school teaching, he turned his attention to Mississippian palaeobotany. In the Scottish Borders, where he then lived, there had been some earlier work on petrified material by William Gordon (1884-1950), Professor of Mineralogy at King's College, London. Long returned to some of the old localities, as well as discovering many new ones, and found beautifully preserved material that had never been described. In a series of classic papers, he described many simple pteridosperm seeds with some of them still in their rudimentary cupules (see Chapter 8). Long also turned to the study of the fossil tree *Pitys* (first described by Witham in 1833), which was widely interpreted as an early conifer. Long was able to show that it was in fact a pteridosperm with fern-like foliage, seeds in cupules, and pollen organs.

Coal petrology and palynology

The study of coal itself became an important issue because of the differing qualities, and therefore value, of the coal recovered from different seams. However, as it is the fossilised remains of peat, the composition of coal also has potential palaeobotanical and palaeoecological interest. One of the pioneers in its study was Marie Stopes (1880-1958; Fig. 2.7) who, with Richard Vernon Wheeler (1883-1939), published in 1919 an account of the constitution of coal in which they originated the terms clarain, durain, fusain and vitrain for the different petrographic types of coal. A modified form of their classification is still in use today. Marie Stopes had first researched palaeobotany in London and then Manchester, and had travelled widely to visit other palaeobotanists, even visiting Japan, before she became interested in coal (see also Stopes, 1919). Then in 1918 she also published her controversial book *Married Love*, and in 1921 opened the first scientific birth control clinic in the world. She had thrown herself headlong into controversy and from then on she had little time for plant fossils, although she continued to attend scientific meetings for the rest of her life. Marie Stopes pioneered a social revolution,

Fig. 2.7 Marie Stopes (1880–1958). Reproduced with permission from Marie Stopes International UK organisation.

while writing books, plays, poems and political pamphlets. She was certainly one of the most colourful palaeobotanists of all time.

It was also of great importance to be able to use coal samples for identifying coal seams discovered in prospecting boreholes, especially in areas of severe underground faulting. From this need emerged the study of spores prepared from the coal and from the sediments associated with them. Henry Witham is accepted as being the first person to figure spores seen in a thin section of coal. P. F. Reinsch published the first attempt at an artificial classification of microspores isolated from coal in 1884, while James Bennie and Robert Kidston gave the first excellent descriptions of megaspores in 1886. Further work was slow to come, however, because there appeared to be no interest or value in it, but in the 1930s, a number of Europeans started to publish accounts of dispersed spores and the literature on this topic gradually grew. In 1933 Ahmet Can Ibrahim (1900–1981) published a morphological classification of spores and used a binomial system of nomenclature that is still basically in use today. Another pioneer was Arthur Raistrick (1896–1991) who published a number of papers between 1932 and 1940. His 1934 paper correlating coal seams in Northumberland was met with great enthusiasm and hailed as a novel method with high promise for the future.

The first monographic works attempting to bring order to this rapidly expanding subject came from the Illinois Geological Survey in the USA. James Schopf (1911–1978), Leonard Wilson (1906–1988) and R. Bentall in 1944 and Robert Kosanke (1917–1996) in 1950 put the relatively new subject of palynology firmly on a scientific footing and their publications instantly became the standard reference works. Palynologists around the world started to specialise either on microspores or megaspores. The Dutch led the way in Europe in megaspore studies, with Sijben Jan Dijkstra (1906–1982) who worked at Heerlen with Jongmans. Dijkstra published a great deal on Carboniferous megaspores from 1946 onwards and much of it was on foreign material, either brought back by Jongmans or sent by Jongmans' overseas contacts. In 1955 the Germans Robert Potonié (1889–1974) and Gerhard Kremp published the first of a series of monographs on spores from the Ruhr Coalfields, which again became standard works. However, megaspores have never really proved to be very good stratigraphical markers, possibly because of the tendency of researchers to define very broad species and to synonymise other authors' species into their own.

Microspores have proved to be much better stratigraphical markers and are used by many more palynologists. There is not the same tendency to lump species in microspore research and, indeed, there is often a tendency to over-divide species or create new ones. A landmark publication came in 1967 from the British National Coal Board with Harold Smith and Mavis Butterworth's (1927–1996) work on microspores from British coal seams (incidentally, this was the first of the splendid series of special publications by the *Palaeontological Association*). Starting from its use in Pennsylvanian stratigraphy, palynology was seen to be of stratigraphical use for rocks of other periods. Now it has become an essential tool in exploration work, most notably in that of the petroleum industry.

For many years the study of palynology was regarded as the reserve of stratigraphers, with little direct interest to the palaeobotanist. This was partly due to the uncertainty over which plants had produced the various types of spore and pollen that were being recovered from the rocks, especially of Palaeozoic age. However, this problem has been progressively overcome as more reproductive structures have been investigated and the *in situ* pollen or spores extracted. Apart from Göppert's account of pollen recovered from a well-preserved alder catkin, however, progress remained slow until the late 1940s, when the economic importance of palynology was at least partly responsible for encouraging the investigation of *in situ* pollen and spores. Robert Potonié compiled the available information in his *Synopsis der Sporae in situ*, the first part of which was published in 1962, and more recently, in 1995, Basil Balme has produced an updated catalogue in the *Review of Palaeobotany and Palynology*.

Using this knowledge about correlating plants with particular types of pollen or spores, it became possible to use assemblages of dispersed spores for floristic and ecological studies. In the early 1960s, Harold Smith, working for the British Coal Board, set about interpreting Carboniferous palaeoecology using dispersed spore data to illustrate a changing spore assemblage that he correlated with coal petrography essentially using Marie Stopes's four coal types. He interpreted the different types of spore assemblages as reflecting a changing plant assemblage from a closed lepidodendroid-dominated swamp to an open moorland-type of community dominated by smaller lycophytes. The swamp was surrounded by drier land on which cordaites and conifers grew. Although current ideas differ in detail from Smith's model, his work produced a significant change in our view of the Coal Measures forests; no longer were they seen as a fairly homogenous flora, but rather as a complex mosaic of plant communities.

Much of this work focussed on the palynology of the coals, which can give a very detailed picture of the local vegetation. More recently attention has been turning to the palynology of the clastic deposits between the coals. Because these clastic deposits were formed during flooding events that resulted in open areas of standing water within the forests, their pollen and spores tend to give a much more representative sample of the regional vegetation. Such an approach obviously loses much of the fine ecological detail available from coal palynology, but has the advantage of allowing better insights into the regional trends in vegetation-change of the forests. Moreover, the clastic deposits also include rare 'exotic' pollen and spores blown in from more upland vegetation, which is rarely, if ever, represented in the macrofossil record. This has, for instance, allowed us to see evidence of plants such as cycads and conifers (see Chapter 9) of Pennsylvanian age, much older than the earliest macrofossils of such plants.

The *Glossopteris* flora and continental drift

We now know that during very late Palaeozoic and much of Mesozoic times, all the world's major landmasses were joined together into one super continent called Pangaea (see Chapter 11). Crustal movements starting essentially in Cretaceous times split this landmass up and moved the fragments apart to give today's continents. This means that the geographical distribution of the different types of past vegetation preserved in the fossil record often seem to make little sense when viewed in the context of today's geography.

One notable discrepancy is the widespread distribution in Permian rocks of a distinctive set of plants (mainly trees) with leaves called *Glossopteris* (see Chapter 8). Leaves of these plants from India and Australia were first described and named by Brongniart in 1828 (in his *Prodrome d'une Histoire des Végétaux Fossiles*) and similar fossils were later found in South Africa in 1867 and South America in 1870. These places are today in relatively low latitudes and, except for the wide oceans that separate them, the distribution of *Glossopteris* is perhaps not impossible to explain in terms of today's geography. However, problems started to arise when it was discovered that the leaves are often associated with glacial deposits that make little sense in the context of their current latitudes. The concept of a *Glossopteris* flora was introduced by Georg von Neumayer in 1887 in his first volume of *Erdgeschichte*. The situation appeared to be further complicated in the early twentieth century, when very similar leaves were discovered in Antarctica by Captain Scott during his fateful second Polar Expedition in 1911/12. Specimens collected by the expedition at Mount Buckley at 85° S were among the rock and fossil samples found in Scott's tent and brought back to Britain in their ship *Terra Nova*.

Albert Seward (1863–1941; Fig. 2.8) described Scott's plant fossils in 1914, when he also appended his views on the characteristics of the southern flora, including their frequent association with glacial deposits. He dismissed one notion that was favoured by some geologists of the time – that the Earth had shifted its axis of rotation. Instead, he suggested that there had been a considerable change in global climates, although the problem of a long polar night clearly worried him. He believed that Scott's

Fig. 2.8 Sir Albert Charles Seward (1863–1941). From Obituary Notices of Fellows of the Royal Society (1941).

new evidence supported the view that had been advanced by other palaeobotanists that an Antarctic continent was the original area of development of many plant groups.

The currently accepted explanation was proposed not long after Seward's paper. Alfred Wegener (1880–1930) had noted that there were many apparent discrepancies in the distribution of fossils, although the *Glossopteris* flora was one of the most marked, and in 1915 (in his book *Die Entstehung der Kontinente und Ozeane*) he came up with the novel explanation that the position of the continents was not fixed. Wegener was not a geologist (he was a meteorologist by profession) and most of the geological community rather dismissed his ideas. It was not for another two decades that continental drift, as it has come to be known, started to be taken seriously, largely through the work of the South African geologist Alexander du Toit (1878–1948), who again noted the scattered distribution of the *Glossopteris* flora. In his 1937 book *The Wandering Continents*, du Toit

proposed that in Late Palaeozoic times, South America, South Africa, Australasia, Antarctica and India were fused together in a large continental mass, which he named Gondwanaland. Most geologists continued to dismiss these apparently preposterous ideas, largely because nobody could come up with an explanation of how continents could move about in this way. Eventually, however, evidence of the mechanism now generally referred to as plate tectonics was discovered in the 1960s. Most scientists now accept this as being the underlying explanation for the distributional patterns of continents on the Earth's surface.

Work on the *Glossopteris* flora was carried on by a number of palaeobotanists, especially in India and South Africa. One of the most important of the Indian palaeobotanists was Birbal Sahni (1891–1949; Fig. 2.9a) whose interest in plant fossils came from studying under Seward in Cambridge. In 1921 he became the first Professor of Botany in Lucknow, but more importantly was one of the founders of India's Palaeobotanical Society in 1946. Later the same year the Society established the Institute of Palaeobotany with Sahni as its first Honorary Director. Unfortunately, Sahni died a week after the Foundation Stone of the building was laid, and the Institute was named after him. Another research school for palaeobotany and plant morphology was established by Divya Darshan Pant (1919–2001; Fig. 2.9b) at Allahabad, who continued work on the *Glossopteris* flora into the second half of the twentieth century.

Continental drift and plate tectonics have also helped to explain the present distribution of other fossil floras. The coal-bearing deposits found in Pennsylvanian and Lower Permian deposits form a distinct band stretching from western North America, through Europe and parts of central Asia, to China. The plant fossils associated with these coals suggest vegetation of a tropical character, but their present-day distribution is largely in northern temperate latitudes. As shown in Chapter 11, however, if we plot these deposits on a map showing the positions of the continental plates as they were in Late Palaeozoic times, we find them occurring mostly within the tropical belt of that time.

Fig. 2.9 Palaeobotanists in India. a. Birbal Sahni (1891–1949). b. Divya Darshan Pant (left) with William S. ('Bill') Lacey, a British palaeobotanist, who also published on Gondwana plant fossils, at the 1975 Leningrad International Botanical Congress (St Petersburg). Photo by W. G. Chaloner.

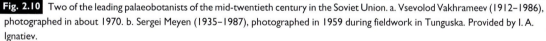

Fig. 2.10 Two of the leading palaeobotanists of the mid-twentieth century in the Soviet Union. a. Vsevolod Vakhrameev (1912–1986), photographed in about 1970. b. Sergei Meyen (1935–1987), photographed in 1959 during fieldwork in Tunguska. Provided by I. A. Ignatiev.

Another distinct set of floras of this age has been found in Siberia and northern Asia, and is known as the Angaran flora. This flora has been studied since the 1840s but it was the Russian palaeobotanist Sergei Meyen (1935–1987; Fig. 2.10b) who showed it to consist of mainly archaic-looking plants, that probably had very limited or no contribution to the world's later Mesozoic floras. When these Angaran floras are plotted out on Late Palaeozoic palaeogeographical maps, they occur in northern high palaeolatitudes.

The Russian school of palaeobotanists under Vsevolod Vakhrameev (1912–1986; Fig. 2.10a) led the way in the investigation of past plant biogeography from the fossil record. They compiled vast data-sets of plant distributions, especially of Palaeozoic and Mesozoic age, which were summarised in their book *Paleozoiskie i Mesozoiskie flori evrazii i fitogeografiya etogo vremeni* (1970). This was unfortunately never translated into English, although a German translation (by Rudolf Daber) was published in 1978, and a summary is given in Meyen's 1987 book *Fundamentals of Palaeobotany*.

Early land plants

The first person to give any clear understanding of early land plants was the Canadian geologist John William Dawson (1820–1899). After studying in Edinburgh, Dawson became Superintendent of Education for Nova Scotia, Canada. Through his extensive travels he was able to study geology at first hand, which led in 1855 to his monumental book, *Acadian Geology*. Dawson then became principal of the fledgling McGill University. A visit by Sir William Logan, head of the Canadian Geological Survey, encouraged Dawson to study the Devonian plant fossils found along the Gaspé coast in New Brunswick. Most of Logan's own extensive collection was lost in a shipwreck, but the few that Dawson saw led him to visit Gaspé to collect more. In 1859, he published his account of *Prototaxites* and *Psilophyton* (see Chapter 4). Other publications by Dawson on Devonian plants followed, including his *Geological History of Plants* in 1888, but they all received little attention. European palaeobotanists interested in the origin of plants just ignored his work, preferring to theorise and base their ideas on plants living today. Their entrenched and introverted ideas held back palaeobotanical studies on early land plants for decades.

In 1913, Dr William Mackie of Elgin reported how he had accidentally found the now famous Devonian chert at Rhynie in Scotland. He initially found fragments of the chert in a dry stone wall near Rhynie village and these contained numerous round structures, initially thought to be volcanogenic vesicles. However, when examined in thin section under the microscope, they were found to be anatomically preserved stems of the plant we now know as *Rhynia*. The Scottish Geological Survey then made a series of excavations in the neighbouring field, and discovered the chert *in situ*. The resulting specimens were sent to Robert Kidston and, in the natural course of events, Kidston would have studied them with Gwynne-Vaughan. Instead, it was Gwynne-Vaughan's friend and ex-colleague at Edinburgh, William Lang (1874–1960), then Professor of Cryptogamic Botany at Manchester, who worked on the chert with Kidston. Their anatomical descriptions of these simple land plants were published between 1917 and 1921, and it is perhaps not surprising that their first description was of a plant they called *Rhynia gwynne-vaughanii*, and a later one named as *Asteroxylon mackiei*. Their work on the Rhynie Chert not only vindicated Dawson's much earlier work, but persuaded Lang to make further studies on early land plants and encouraged other palaeobotanists to look elsewhere for remains in Devonian and older rocks (see Chapter 4). After the original Rhynie Chert excavations, the exposures were covered over and the field returned to agricultural use. However, there have been a number of subsequent excavations (e.g. Fig. 2.11) such that there is now a substantial amount of material available in various institutions around the world.

Further collecting along the New Brunswick coast was soon undertaken by Loren Petry (1887–1970), Professor of Botany at Cornell University. He was followed in this by other, more productive palaeobotanists like Harlan Banks, Chester Arnold (1901–1977: Fig. 2.12) and Fran Heuber who have described a wealth of new and important material from this site – a hundred years or so after Dawson's original paper.

With the probable exception of the Australian *Baragwanathia* flora, most of these discoveries were in rocks of Devonian age. However, the evidence was clearly pointing to land plants having first evolved in Silurian or even older floras. In 1937, Lang established the genus *Cooksonia* from Late Silurian strata in the English/Welsh borders (it was named after Isabel Cookson, with whom he had worked earlier on the *Baragwanathia* flora). As will be shown in Chapter 4, this is a small and extremely simple organism and it proved very difficult to confirm that it was in fact a true land plant; to do this, it

Fig. 2.11 Collecting in the Rhynie Chert trench cut for the International Botanical Congress in 1964 (left to right: A. Lyons, W. Lacey, J. Richardson (hidden), J. Vigran and W. G. Chaloner). Photo by B. A. Thomas.

Fig. 2.12 Chester Arnold (1901–1977). Photo by K. McCandlish, provided by E. L. Zodrow.

using an array of new techniques, were finally able to prove conclusively that the Late Silurian *Cooksonia* was a vascular land plant. The search is now on for even earlier evidence of land plants, with new palynological data suggesting possible origins in Early Silurian or even Ordovician times.

The age of cycads

Research on Mesozoic floras was for many years overshadowed by the overwhelming interest in Late Palaeozoic floras – party because of the economic importance of the latter's coal deposits, and partly because Palaeozoic plants seemed to be so different from today's plants. Nevertheless, there were people who collected plants from Mesozoic rocks, and many of these were naturalists with much broader interests who combined their study of plant fossils with other aspects of natural history and geology.

In the early nineteenth century, amateur and professional collectors were at work on the Yorkshire coast where they were collecting Jurassic adpressions. The most notable were William Bean and John Williamson (the father of William Williamson referred to earlier). Brongniart also described and figured 22 species from Yorkshire, while Lindley and Hutton included 43 of them in their publication.

was necessary to demonstrate the presence of a number of key morphological characters, notably a vascular strand and stomata. This evidence remained elusive until the 1980s, when Dianne Edwards and her colleagues at Cardiff University,

William Buckland (1784–1856) was a priest, Fellow of Corpus Christi College Oxford and later Reader of Mineralogy there. Eventually, in 1845, he became Dean of Westminster. He was an avid 'Diluvianist', his most notable contribution to this field being his 1836 book in the *Bridgewater Treatise* series, in which he tried to explain 'The Power, Wisdom and Goodness of God as manifested in the Creation' through his geological observations. Buckland was mainly interested in animal fossils, and is perhaps best known for having found some of the earliest-discovered dinosaur remains. His most important palaeobotanical contribution was to interpret the silicified stems that were being found in the quarries at Portland Bill in Dorset as cycad-like stems and in 1828 Buckland named them *Cycadeoidea*. The local quarrymen were also finding ring structures that they called 'birds' nests' (Fig. 9.10). These have subsequently been interpreted as stromatolite rings that were formed around the bases of conifer stumps that had subsequently rotted away. Buckland also collected Mesozoic plants from the Middle Jurassic Stonesfield Slate in Oxfordshire. Buckland himself published little on this flora, but he arranged for paintings of many of the fossils to be sent to Sternberg in Prague, who included a number in his book *Flora der Vorwelt*. As a consequence, for many years the Stonesfield flora was widely regarded as the epitome of Jurassic vegetation.

William Carruthers (1830–1922) was Assistant Keeper and then Keeper of Botany in the British Museum (Natural History). Although his work was mainly on living plants, he did publish a number of accounts on cycads and bennettitaleans (see Chapter 9 for a discussion on these plant groups) from the Yorkshire Jurassic, establishing the name *Beania* for a cycad seed. Carruthers named this seed after William Bean who had earlier sold a large collection of plant fossils to the Museum.

Interest in Mesozoic palaeobotany advanced substantially during the last years of the nineteenth century, largely through the efforts of Albert Seward. Seward spent a year studying with Williamson in 1886, focussing largely on Carboniferous floras, and then spent time travelling in Europe, looking at many of the major palaeobotanical collections there. However, after being appointed Lecturer in Botany at Cambridge in 1890 (he eventually became Professor of Botany there, and then Master of Downing College) he switched his interests to Mesozoic fossils. Arguably his most important palaeobotanical work is summarised in a series of catalogues that he produced of the British Museum (Natural History) collections of Wealden plants between 1894 and 1895, and Jurassic plants between 1900 and 1904. These did not merely list the specimens, but included detailed descriptions, illustrations and interpretations of the fossils, and for the first time brought together what was known about Mesozoic floras, notably the Middle Jurassic floras of Yorkshire and Oxfordshire, and the Early Cretaceous floras of south east England. These publications also reflect a rather amazing aspect of Seward's research: he hardly ever collected any specimens himself, and relied almost exclusively on museum collections. Indeed, Professor Tom Harris, one of Seward's students, told one of us (BAT) as one of his students, that Seward had only collected one British specimen in his life and his only collecting trip was to Greenland when he was accompanied by Richard Holttum (1895–1990; who went on to become a world authority on living ferns). However, despite this lack of field experience and the fact that he never saw the need to prepare cuticles, Seward became arguably the most famous palaeobotanist of his time, a reputation at least partly based on his two great published works: the encyclopaedic, four volume *Fossil Plants* (1898–1919) and the popular *Plant Life Through the Ages* (1933) which gives a readable account of the subject.

Hugh Hamshaw Thomas (1885–1962; Fig. 2.13a; see Harris, 1963) was also at Cambridge for most of his career. He first worked with Newell Arber on Carboniferous floras, and through him met Robert Kidston, who taught him the importance of field collecting. In 1911, Thomas visited Stockholm where he met Alfred Nathorst (1850–1921), who had just refined the maceration techniques for extracting cuticles from fossils. This visit proved pivotal for the development of Thomas's future research. Although fossil cuticles had been noticed since the early nineteenth century and were first recorded in 1834 by Lindley & Hutton, their separation from the fossil often

a.

b.

Fig. 2.13 British Mesozoic palaeobotanists. a. Hugh Hamshaw Thomas (1885–1962). From the Royal Society (London). b. Thomas ('Tom') Maxwell Harris (1903–1983), photographed at Philpots Quarry, Sussex, England.

proved difficult so their study had not really progressed. Thomas encouraged one of his students at Cambridge, Lucy Wills (1888–1964) to study Carboniferous cuticles, but this work never properly developed since she published only one paper, in 1914, before pursuing a medical career. Instead, Thomas was drawn to change direction in his research by another palaeobotanist whom he met in Stockholm, Thore Halle (1884–1964). Halle had earlier visited the Yorkshire coast and had rediscovered many of the old plant fossil localities that were believed to be lost or worked out. This inspired Thomas to go to Yorkshire, but he found that collecting there was not easy. He sought-out the famous Gristhope locality, but initially had no luck and concluded that it was lost. Deciding instead to collect seaweed, he tore up a stubborn piece attached to a rock, and there were the fossils. This amazing chance find provided him with specimens for his most important work, where he was able to show that the seed-bearing *Caytonia*, the pollen-bearing *Antholithius* and the leaves *Sagenopteris* were parts of the same plant (see Chapter 9). Although Thomas's original conclusion that these plants were angiosperms has proved mistaken, the Caytoniales as they have become known, have proved central to many of the subsequent discussions on gymnosperm evolution.

The next major work on the Yorkshire Jurassic flora was by Thomas ('Tom') Maxwell Harris (1903–1983). Harris had intended to become a country doctor, but turned to botany instead. He was an undergraduate at Cambridge when Seward was Professor of Botany, and Hamshaw Thomas was teaching palaeobotany. However, Harris's decision to enter palaeobotany was attributed to lectures from Henry Smith Holden (1887–1963) at Nottingham rather than those from Hamshaw Thomas at Cambridge, which he found soporific (Holden published only a few palaeobotanical studies, and, in fact, later became better known as a forensic scientist, eventually running the Forensic Science Laboratory of the London Metropolitan Police). Harris also coincided briefly with John Walton at Cambridge; although Walton soon moved to Glasgow to become Professor of Botany. Harris first worked on the Rhaeto-Liassic flora of Greenland that had been sent to Seward, and then spent winter there with the Greenland Geological Survey expedition to collect more material. This formed the basis of his first major work published between 1931 and 1937, which was considered to be so good that in 1935 he was selected for the Chair of Botany at Reading, where he stayed for the rest of his career. Harris learned the techniques of preparing and studying cuticles that would

Fig. 2.14 Views across the Petrified Forest National Park, Arizona, USA. a. Broken logs in the Rainbow Forest in 1935. b. A balanced log in the Rainbow Forest in about 1950. c. The Agate Bridge in the Jasper Forest in about 1908. Photos courtesy of the US National Park Service (a, b), American Memory, Library of Congress (c).

form the basis of much of his future work from Halle at Stockholm. The Yorkshire Jurassic was taken up by Harris after moving to Reading and he told one of us (BAT) that he had 'stolen it from Hamshaw Thomas'. In 1900, Seward believed that he had described everything of value from the Yorkshire Jurassic and that there was probably nothing worthwhile left to collect. In essence, Seward had carried on his researches in the same tradition as Sternberg and Brongniart, and although he had once prepared a cuticle decided that they were of no taxonomic value. Hamshaw Thomas had clearly shown this to be incorrect, but it was Harris that really showed the value of cuticles in a long series of papers and in his five-volume *Yorkshire Jurassic Flora* (1961–1979). Largely through Harris's work, the Yorkshire Jurassic flora is now regarded throughout the world as the classic of fossil flora of this age.

Although he was helped in his work by the collections from the Yorkshire coast that had been amassed by Maurice Wonnacott (1902–1990) of the British Museum (Natural History) Harris collected much of his material himself. After the end of the World War II, Harris visited Yorkshire at least once a year, usually on holiday with his family, but sometimes with other palaeobotanists. One of us (BAT) experienced collecting with Harris, and found it both interesting and instructive. He always maintained that there was more to collect and research, and that it is usually the collector and not the locality that becomes exhausted. More recently work on the Yorkshire Jurassic flora has been published by Chris Hill and Han van Konnigenburg-van Cittert.

In America, Mesozoic plants of a completely different kind were being discovered and they were exciting the public's imagination as never before. In 1849, Lieutenant Simpson of the United States Army Corps of Topographical Engineers was accompanying a military expedition into Navajo country in northeastern Arizona, when they found 'petrified trees' in what were later shown to be Triassic rocks of the Chinle Formation. Other US Army units later found further large masses of 'petrified trees' in what was subsequently named *Lithodendron* Wash. The trees, identified as conifers, were described by Simpson and others, and as a result collectors started to come in increasing numbers (Fig. 2.14). Souvenir hunters and commercial enterprises were soon removing large quantities of material and gem collectors were blowing them up to recover crystals from cavities in the trunks. The local population started to complain and, when in the early 1880s, a mill was built to crush the petrified wood into abrasives, local indignation boiled over. As a result the Territorial Legislature (Arizona was not yet a state) petitioned Congress to protect the area. In 1900 collecting was prohibited and in 1906 some of the area was declared a National Monument. This was the first plant fossil site to receive official state protection. It was subsequently enlarged and on 8 December 1962 was declared the Petrified Forest National Park. In 2004, President George W. Bush signed the Petrified Forest National Park Expansion Act adding another 125 000 acres, more than double its size.

The Black Hills of Dakota yielded further petrified material in 1893, but this time it was fossil bennettitaleans that were found (see Chapter 9). They were studied by George Reber Wieland (1865–1953), who purchased and collected hundreds of specimens that were taken to Yale University. Wieland's work clarified many details, especially the structure of their reproductive organs. His work was continued by Theodore Delevoryas and William Crepet.

The third important site for petrified material in the Americas is far away from the other two, in Patagonia in southern Argentina. Here there are petrified logs and cones of late Triassic age that once again became collectors' items soon after they were described in the 1920s. Longitudinally cut and polished cones were favourites; many museums soon had large collections and examples can be found in many university teaching collections. Several people, including Wieland, Mary Calder (1906–1992), Ruolf Florin (1894–1965) and more recently Ruth Stockey have researched and published details of them.

Flowering plants

The origin of the angiosperms was regarded as a complete mystery for far too long. 'Ancestral angiosperms' were sought in the Jurassic and Cretaceous floras, but none was found. Fossils of true angiosperms were seen as suddenly appearing in the fossil record, persuading many people to speculate about evolution in the uplands where fossilisation was unlikely if not impossible. This belief that plant fossils would supply no effective answer led to many papers being written in the early part of the twentieth century on the theoretical aspects of angiosperm evolution. The abandonment of one idea to be followed by another led palaeobotany into disrepute in many people's eyes, and plant morphologists sought their answers in comparative morphological studies of living plants. In the last thirty or so years, however, the balance has been redressed, and we now know much about the evolutionary changes that were occurring within the early angiosperms, even if we do not know the precise group from which they originally evolved.

Angiosperms shed their organs much more readily than gymnosperms and spore-bearing plants. Several calculations have been made of the number of the various organs that are shed or abscissed during the life of a large woody angiosperm. A valuable estimate made in 1976 by Norman Hughes of the relative quantities and time available for fossilisation of angiosperm organs from a single tree quotes: 10^7 pollen grains shed in 2 months every year, 10^3 flowers shed in 1 month every year, 10^3 seeds shed in 2 months every year, 10^5 leaves shed in 4 months every year, 10^2 twigs shed for 2 months every year, large woody stems and roots shed only once every 10 years (Fig. 10.4).

Obviously not all abscissed organs become fossilised. Leaves are by far the most common flowering plant macrofossils found by collectors, although pollen grains can be recovered in enormous numbers from rock samples. However, the first flowering plant fossils to catch the imagination of the early naturalist-collectors were not leaves, but seeds and fruits that were found pyritised in the Palaeogene London Clay of southern England. In the early nineteenth century, locals were collecting pyrite nodules and clay iron-stone for sale, scouring these deposits along the northern shore of the Isle of Sheppey in the Thames Estuary. They also found pyritised seeds, fruits, twigs, fish, crabs, shells and sharks' teeth, which they sold to tourists who came down from London on the river steamers.

In 1757, James Parsons (1705–1770) described and figured some of these seeds and fruits, and then on the basis of their ripeness deduced that the biblical flood had occurred in the autumn. Over fifty years later, in 1810, Francis Crow described more than 700 species that he had amassed over twenty years and concluded that they once belonged to a tropical or high southern latitude vegetation; quite a far-sighted interpretation considering this was the beginning of the nineteenth century. Finally, in 1841, James Scott Bowerbank (1799–1877) produced his monumental *A History of the Fossil Fruits & Seeds of the London Clay*. Bowerbank was a well-off amateur palaeontologist who formed the London Clay Club with a number of other enthusiasts. This Club gave rise to the Palaeontographical Society, which is still producing monographs on both plant and animal fossils.

Little further serious work was done on the flora for many years until Mary Reid (1860–1953) and Marjory Chandler (1897–1983) formed their partnership in 1920, which remained unbroken until Reid's death. Together they produced the monumental *London Clay Flora* in 1933, followed in 1961 and 1973 by Chandler's supplements. The work was recently taken up by Margaret Collinson, who has also written a field guide to the flora that was published by the *Palaeontological Association* in 1983.

Other seed and fruit floras were found and studied on the continent from the beginning of the nineteenth century, and both Schlotheim and Sternberg knew about them. However, there were no studies of similar detail to those carried out in England until the middle of the twentieth century. Similarly in the USA, studies were spasmodic until as late as the 1950s, when comprehensive studies commenced on the Tertiary Brandon lignite of Vermont.

Leaves of flowering plants are now known from the Cretaceous onwards. For many years it was thought that even the oldest known leaves could be referred to modern families and genera, and that they could not provide any evidence about angiosperm evolution. It is only since the early 1970s that more critical studies on leaf architecture and on associated pollen has shown this to be a false premise. Instead, we now know that there was a major Cretaceous adaptive radiation of flowering plants between about 130 and 90 million years ago. Recent studies in North America show that angiosperms appeared in the eastern coastal plains in late Barremian or early Aptian times, and spread rapidly westwards and northwards.

Studies on leaf shape and venation patterns can also be used to estimate palaeoclimatic limits. The first approach in the 1960s was to use the principle of 'the nearest living relative', that is to make palaeobiological assumptions based upon the environmental parameters of the living taxa. This approach proved difficult to apply because it relied on accurate identifications of the fossils and critical comparisons with living species. In the 1970s, a different approach was developed that used 'foliar physiognomy', which matched leaf types rather than relying on species identifications. The eight characters used are: leaf size distribution, leaf margin type, drip tips, organisation (i.e. simple or compound), venation-pattern, venation density, leaf texture and leaf base type.

Fossilised flowers were virtually unknown until fairly recently. It was not until intense collecting commenced in the 1970s that a spate of flowers were found and described in the literature. Many flowers have now been described either as adpressions or as fusainised three-dimensional remains, with the most extensive collections coming from western Portugal and eastern North America. Another recent parallel development has been the study of pollen grains obtained from the stamens of flowers. This has given important systematic information on pollen types that have been previously recognised only as dispersed grains, some of which were shown by Else Marie Friis and Kaj Pedersen in 1996 to be stratigraphically important.

Phylogenetic analysis (cladistics) has been widely used to hypothesise on many aspects of plant and animal evolution, but is consistently used to investigate the origin and early diversification of angiosperms. Using this technique, Peter Crane, Else Marie Friis and Kaj Pedersen challenged the generally accepted view that the earliest angiosperms were *Magnolia*-like, suggesting instead that they were perhaps herbaceous plants with simple, unisexual flowers lacking differentiation of sepals and petals. There is clearly a great deal left to be considered, and future collecting and research is needed, especially on Early Cretaceous floras from low palaeolatitudes.

Extinction and disruption at the Cretaceous/ Tertiary (K/T) boundary have dominated the research of a number of palaeontologists in recent years. One idea, based on the discovery of a distinctive layer of iridium very near the K/T boundary in the Yucatan Peninsular of Mexico, was that an asteroid impact was responsible for the extinction of the dinosaurs and other groups of large reptiles such as the pterosaurs and pleisiosaurs. For many years the idea of an asteroid having killed-off the dinosaurs became fixed in the public mind. More recently, however, this link has been questioned and it has been suggested that large-scale volcanic activity in India, which resulted in the formation of the Deccan Traps, may instead have caused the extinction. The subject is still highly controversial, with both schools of thought having their supporters, as well as groups suggesting that it was a combination of the events that caused the disruption, and others suggesting that these groups of animals were just dying out anyway. Whatever the explanation for the extinction of these animal groups, the K/T extinction was less profound for plant life. Some recent work has concentrated on the development of vegetation after the event. For example, in 1993 Gary Upchurch and Jack

Wolfe used leaf analysis to recognise two major changes at the K/T boundary: a major increase in precipitation and the development of a zonal distinction between the broad-leaved evergreen vegetation of the southern Rocky Mountains/ Gulf Coastal Plain and the broad-leaved deciduous vegetation of the northern Rocky Mountains. The consequences of the K/T event for plant life are discussed further in Chapter 11.

The future for palaeobotany

As we have tried to show, palaeobotanists have studied plant fossils for many different reasons. The very first were naturalists in the broad sense who were interested in the variety of life on Earth that they saw around them; plant fossils were investigated to see how plant life in the past compared with that of today. Then, in the early nineteenth century, the study of fossils became linked to geological investigations, especially the understanding of sedimentary rocks and their sequence in time.

These two different aspects of palaeobotanical study remain with us today, but the differences between them have often become blurred through studies of evolution, palaeoecology and palaeogeography. Furthermore, the acceptance of continental drift as a reality involving the movements of crustal plates has opened the door to a better understanding of plant distribution and speciation. Our knowledge of palaeoecology, and the interactions between animals and plants, has increased tremendously in recent years and shows no signs of decreasing in importance for future research work. Similarly, reproductive biology is an ever expanding research field that will continue to give us a better insight into whole plant biology.

New techniques introduced in recent years have included biochemical analysis and the use of cladistics as aids to recognising relationships between taxa. Both have given useful and interesting results, and their use will undoubtedly increase as new discoveries of plant fossils are made, but where they may lead us in the future is still uncertain. Electron microscopy, both scanning and transmission, has allowed palaeobotanists to look at ever finer detail of the fossils, such as the fine surface-features of cuticles and the ultrastructure of pollen exine. Computer modelling has allowed the mechanical strength of various parts of extinct plants to be estimated, which can throw light on how they originally lived.

Palaeobotanists are also using experiments on sedimentation to simulate fossilisation in an attempt to understand how the fossil assemblages may relate to original plant communities. From this can come environmental interpretations. Such work is naturally fraught with difficulty because the further back in time we go the less certain we are of our interpretations. Even something seemingly simple like temperature regimes becomes more complicated when latitude and overall global temperature changes are taken into account. Much more integration of different disciplines will be essential if further progress is to be made.

Information can be hard to find even in this age of fast communication. The literature is ever expanding and the number of journals increasing. There have been attempts to alleviate this problem. The *Fossilium Catalogus* was started in 1913 by Wilhelmus Jongmans and edited by him until his death in 1957. This aims to list every published record of every species of plant fossil in the world, and is still continuing to be published today (Part 108 was published in 2006). Although inevitably incomplete, it is the most comprehensive encyclopaedia of palaeobotany that we have available and is of immense value for most groups of plants, perhaps with the notable exception of the angiosperms. Other attempts at compiling published records were made by Tralau (1974), Tralau & Lunblad (1983) and Boersma & Broekmeyer (1979–1982). There are also ongoing bibliographies of American (annual) and European (biennial) palaeobotanical literature.

Good field guides are important assets and many more are becoming available. In Britain the Palaeontological Association has published three on plant fossils: Sheppey Tertiary plants by Margaret Collinson in 1983, Carboniferous plants by the present authors in 1994, and Yorkshire Jurassic Plants by Han van Konijnenburg-van Cittert and Helen Morgans in 1998. In North America the most recent guides are on Western North America by Don Tidwell in 1998, the Pennsylvanian of Illinois by James Jennings in 1990, Mazon Creek by Jack Wittry in 2006, West Virginia by Bill Gillespie *et al.* in 1978, and the Pennsylvanian of Nova Scotia by Erwin Zodrow *et al.* in 2001. Such guides are valuable for beginners and amateurs, but it should always be remembered that there is much still to be learnt and these guides are not exhaustive in their contents. Collecting involves selection and the ultimate rejection of the vast majority of specimens deemed to be duplicates or worthless. Rejects regarded as indeterminable (because they are not in the field guides) might be just what someone else is looking for. Collecting, within reason is to be encouraged.

Regular international meetings also help the flow of information. The meetings of the greatest potential interest to palaeobotanists are organised by the International Organization of Palaeobotany (IOP) every four years. There is the International Botanical Congress that meets every six years and provides the venue for decisions on the taxonomic rules that affect the naming of all plants, including fossils. There is also the Carboniferous Congress (*Congrés Internationale de Stratigraphie et de Géologie du Carbonifère*) mentioned earlier.

So, what is the future for palaeobotany? In the last twenty to thirty years there have been many outstanding advances to our knowledge of plant fossils. The origins of land plants and seed-plants remain subjects of great interest among palaeobotanists and, although the broad patterns of their evolutionary history have probably been identified, many details remain to be discovered. The origins of flowering plants remain highly contentious, however, and it is an area which continues to attract interest and controversy. Much effort is being put into trying to reconstruct whole plants from the fossil record, especially as phylogenetic (cladistic) analysis depends on whole organisms to identify patterns of evolutionary history. However, only a very small proportion of fossil plants will ever be fully reconstructed, and the challenge now is to see what particular plant organs such as seeds, leaves, pollen, and so forth, can tell us about plant evolution and environmental change. We have to accept that the fossil record is not complete and can never answer all the questions that we have to ask. Logically, very little plant material is ever preserved because the bulk of the dead plant material must be recycled back into the environment. There again, most previous palaeobotanical work has been focussed on a relatively small part of the Earth's land surface, especially Europe and parts of North America, and as the remoter places are opened-up with easier and safer access, new and exciting discoveries will undoubtedly be made. One only has to look at the major new finds of possible early flowering plants that have been made in recent years in China to realise the potential that new explorations may have. There are undoubtedly many new palaeobotanical discoveries still waiting to be made by anyone who is willing to pursue the subject.

Recommended reading

Andrews (1980), Arber (1921a, 1921b), Bowden *et al* (2005), Burek & Higgs (2007), Gordon (1935), Long (1996), Lyons *et al.* (1995), Meyen (1997), Taylor & Smoot (1984), Thomas (1986), Vakhrameev *et al.* (1978), Ward (1885), Williamson (1896), Zittel (1901).

Chapter 3

Studying plant fossils

As shown in Chapter 1, plant fossils come in many different modes of preservation, each of which has to be investigated in its own way to maximise the information available. This book is not a detailed laboratory manual describing these various techniques; such practical guidance can be found elsewhere, most notably in the book edited by Jones & Rowe published in 1999 by The Geological Society, London. Instead, this chapter gives some background to some of the techniques, suggests when they should be used and what they might reveal.

Morphology of adpressions

An adpression usually first sees the light of day when the palaeobotanist cracks open the rock with his hammer. The rock may fracture along the same plane that the fossil is lying in, especially if it is a flattened organ such as a leaf. If so, the shape of the fossil is revealed. However, if the plant fragment is not totally flattened, or does not lie in the plane along which the rock has split, then it may be necessary to remove some of the rock matrix – a process known as degagement. This can be particularly important when trying to establish details of the branching pattern of stems or the fine structure within compound reproductive structures.

Degagement is best done under a binocular dissecting microscope. Some palaeobotanists favour using an electric engraving tool fitted with a fine tip. This can be useful if the rock matrix is hard, or a large amount of rock needs

to be removed. However, such tools may be too rough for delicate work and can cause damage to the fossil. For fine degagement, the strongest material to use is tungsten wire dipped in molten sodium nitrite, but for most purposes using a small hammer and fine chisels, a dissecting needle, or an old-fashioned steel gramophone needle in a tool-maker's mini-chuck can produce better results. Dental chisels can be useful for such work and can be purchased new or, as we have found, a friendly word with your dentist can yield cast-off chisels that are still perfectly serviceable for palaeobotanical work.

The main drawback with degagement is that it is easy to damage the fossil; if the tool slips and removes some of the fossil or gouges its surface, it can be difficult or impossible to repair the damage. So, such work must be done very carefully. There is also the problem that it can result in ugly white scars in the rock around the fossil, especially those preserved in dark shale or mudstone (Fig. 3.1). This is really only an aesthetic issue but may be a concern if the specimen is to be illustrated in a publication. Some technicians have developed the art of camouflaging these scars with paint of the appropriate shade of grey, but this is a difficult procedure and the results can sometimes look worse than before the remediation was attempted.

Macrophotography

Making a photographic record of plant fossils is central to many palaeobotanical studies, especially if the work is to be published. This is often

Fig. 3.1 The effect of degagement on a specimen of *Mariopteris*, when trying to expose the clamber-hooks at the ends of the pinnae. Coal Measures (Middle Pennsylvanian Series), Cerau Colliery, Maesteg, Wales (David Davies Collection, NMW Specimen 27.110G.1140). Photo by C. J. Cleal.

straightforward, provided there is a reasonable contrast between the fossil and the rock, or the fossil has a fairly good surface topography or texture. A standard macrophotography set-up can be used (film or digital): the camera is fitted with either a macro-lens or stepping-rings, mounted on a stand, and the fossil illuminated with one or more lights, at least some of which should be at a low angle to produce partial shadows to enhance any surface topography. Many modern digital cameras now come with a built-in macro facility that can produce reasonable results, at least for making basic records of specimens.

In many cases, however, there is only a low contrast between the plant fragment and rock, a black fossil on dark-grey shale, for instance, or an impression where the phytoleim has been lost. In such cases it is necessary to try to enhance the contrast and traditionally two approaches have been used in order to make the image clearer. If there is some surface topography, this can be enhanced by coating the fossil with a very thin layer of white powder, such as ammonium chloride. When done well, this can produce dramatic results, but it is very difficult to make an even covering of powder over the fossil. The other main approach is to immerse the fossil in a liquid, to enhance the contrast even if there is no surface topography, but there are even more serious drawbacks with this method. Water can cause

damage to the fine structure of the fossil and the specimen can take a long time to dry afterwards. Organic solvents have often been used instead, but these can be very unpleasant, if not dangerous, to use and might still damage the fossil. Moreover, there can often be problems with back-reflection from the surface of the liquid obscuring parts of the fossil.

It is possible to get over this last problem by fitting a polarising filter to the camera. If the light used to illuminate the fossil is polarised, rotating the polarising filter on the camera lens can dramatically enhance the contrast between fossil and rock. This is a very cheap method that can produce quite spectacular results. Polarised light can be achieved simply by putting a sheet of polarising film over the light source, and a rotating polarising filter for the camera lens can be obtained from most camera shops. The only real drawback with the method is that the polarising film reduces the strength of the light source by about a half, and so either a strong light is needed, or the pictures have to be taken using long exposure-times. Full details of this method can be found in Crabb (2001).

Transfers

A compression fossil normally only shows one surface of the plant fragment and this is often

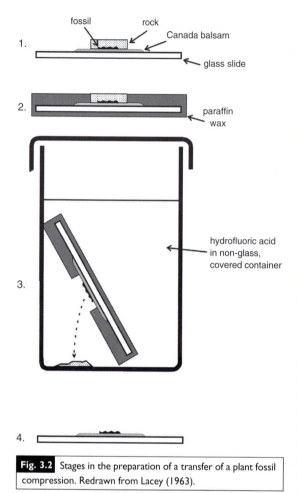

1.

fossil   rock
Canada balsam
glass slide

2.

paraffin
wax

3.

hydrofluoric acid
in non-glass,
covered container

4.

Fig. 3.2 Stages in the preparation of a transfer of a plant fossil compression. Redrawn from Lacey (1963).

microscope slide using a mounting medium such as Canada balsam. To prevent the glass being attacked by the hydrofluoric acid it had to be covered, usually with candle wax. However, if the wax had even the slightest crack in it, the acid would seep in and destroy the glass slide. Nowadays, transfers tend instead to be made using polyester plastics such as 'castoglas', or methacrylate resins, epoxy resins, or polyurethane coating resins, which are not attacked by hydrofluoric acid.

Cuticles and epidermal structures

The study of cuticles has become one of the most important methods of studying plant compression fossils, because these are often the only sources of anatomical information available (Fig. 1.5). The cuticle itself is non-cellular, consisting of a protective waxy layer that covers most aerial surfaces of a plant except where there is bark (periderm). However, because the cuticle extends a little way down the anticlinal walls of the epidermal cells, its inner surface will show the outlines of the underlying epidermal cells, including the stomata, hairs, papillae and glands. By extracting the cuticle, therefore, we can often see details of the epidermis. Some plants have thicker and therefore more easily extracted cuticles than others, and sometimes the whole of the epidermal cells are covered with cuticle revealing the outlines of the underlying hypodermal layer. Others (e.g. many ferns) have very thin cuticles, but even these can sometimes be prepared if sufficient care is taken.

Cuticles are not always preserved in compressions. If the fossil has been too-deeply buried or has been subjected to tectonic activity, the volatile substances in the fossil will have been removed and the cuticle lost. It is normally necessary for at least 28% of the volatiles to remain, which is equivalent to a coalification rank of medium-volatile bituminous coal.

To prepare the cuticle, it is first necessary to remove the carbonaceous phytoleim from the rock. This can sometimes be done mechanically using fine needles, especially if the rock is soft

sufficient for identification purposes. In some cases, however, one surface is much more informative than the other; for example fertile fern pinnae have their sporangia on the lower (abaxial) surface. Inevitably in such fossils, the fracture line of the matrix usually passes along the smoother upper (adaxial) surface of the fossil leaving the informative side face down on the matrix and difficult to see. A technique known as 'transfer' was developed to get over this problem (Fig. 3.2). This involves attaching the fossil face-down onto some medium, and then dissolving the rock away, usually with hydrofluoric acid – a highly toxic reagent that must be used with great care. The fossil is then left attached with the once obscured surface now exposed. In the original method, the fossil was attached to a glass and the fossil relatively robust. Sometimes it may be necessary to dissolve the rock away. With very soft rock this can often be achieved simply by placing the fossil in water but, more typically, more vigorous treatment is needed; for Palaeozoic fossils, treatment with hydrofluoric acid is normally required.

When the phytoleim has been removed, the carbon has to be oxidised away to leave the cuticle. In many Cenozoic fossils this can sometimes be achieved with moderately weak reagents such as hydrogen peroxide, but most specimens require stronger treatment with either concentrated nitric acid alone or Schulze solution (a mixture of nitric acid and sodium or potassium chlorate). This treatment turns the carbon to a brown substance which, after washing, can be dissolved in a dilute alkali solution (usually ammonium or potassium hydroxide). It then needs to be washed in distilled water before mounting. If the cuticle has been successfully extracted from a thin leaf or sporophyll, there will often be cuticle from both surfaces of the plant fragment. It may be necessary to separate the two cuticles whilst they are still in water – sometimes a difficult manoeuvre that is best done in a Petri dish or on a glass microscope slide under a dissecting microscope.

To examine cuticles under a compound optical microscope, the cuticle has to be mounted on a glass slide. Various mounting mediums have been used over the years, but glycerine jelly remains one of the most successful and easy to use. If the cuticles are very thin and details of the epidermal cells difficult to see, the cuticle can be stained. Safranin is the most widely used stain (e.g. Fig. 1.5a) although various others have been investigated by Michael Krings. This can be done before it is mounted. Alternatively, if a little of the stain is dissolved in glycerine jelly, the cuticle will slowly absorb some of the stain after mounting. The outlines of the epidermal cells can also be enhanced optically, using various optical techniques, such as phase and interference-phase (Normarski) contrast (e.g. Figs. 1.5c, 8.10b, 9.14).

Cuticles can also be examined by electron microscopy, which can produce more detailed results than optical microscopy, although the resulting preparation is not as long lasting. For the scanning electron microscope (SEM), which shows details of the cuticle surface, the cuticle is dried onto a small metal stub. It is then usually coated with a thin layer of gold or gold palladium alloy, although some modern SEMs can deal with specimens that have not been coated. The two surfaces of the cuticle will give different information: the inner surface will show the pattern of the underlying epidermal cells; the outer surface will show structures that protrude from the plant, such as hairs, spines and glands. For the transmission electron microscope (TEM), which reveals very fine ultrastructure, the cuticle is embedded in an epoxy resin and then very thin sections are cut with an ultramicrotome.

Epidermal details can sometimes be seen by direct observation using top illumination. The use of incident-light darkfield considerably improves the amount of detail that can be seen (Fig. 3.3) and allows speedier identification of specimens where epidermal detail is a diagnostic character. With some specimens, incident-light darkfield observations can reveal details that were previously unknown and, in some cases, can give useful results after all previous attempts to obtain them have been abandoned.

Extracting *in situ* pollen and spores

Spores and pollen (Fig. 3.4) can be extracted from fructifications by first removing the phytoleim with a mounted needle. These fragments are then macerated as described above for cuticles. Megaspores can sometimes be removed individually before maceration. In contrast, microspores can be liberated in enormous numbers from small fragments of compression. If they stick together in the tube after the final washing, the clumps can be broken up by using an ultrasonic bath. The dispersed microspores can then be concentrated either by centrifuging before mounting, or separated through heavy liquid flotation using such chemicals as stannic chloride or zinc chloride.

If the fructifications are preserved in a coal ball, spores can be liberated with hydrochloric acid before maceration, but if they are in ironstones they may need treatment with

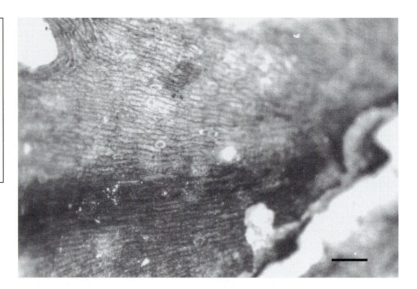

Fig. 3.3 Incident-light darkfield microscopy used to reveal epidermal structure including stomata, of a leaf of *Selaginella gutbieri* (Hirmer) Thomas. Scale bar = 100 μm. Upper Carboniferous, Zwickau Coalfield, Saxony, Germany (Museum für Naturkund, Berlin, Specimen 1980/492). Photo by B. A. Thomas (see Thomas *et al.*, 2004 for further details).

hydrofluoric acid as well as hydrochloric acid to remove both the calcite and silica in the matrix. In these cases megaspores may still retain their three dimensional shape and cannot be mounted in a medium. They are better mounted dry on a slide under a raised cover slip.

Dispersed pollen and spores

There is no single effective technique for extracting palynological samples from all lithological types. Many rock samples will need to be treated with acids to remove carbonates and silicates before maceration with Schulze solution. Cemented rocks such as sandstones can be placed directly into acid treatment but non-cemented materials, such as claystones and siltstones, need to be physically broken up prior to treatment. Further information can be found in palynological texts such as Jansonius & McGregor (1996) and Traverse (2007).

Three-dimensionally preserved plant fossils

Wood is often preserved in the fossil record but other organs are only preserved in three-dimensions in exceptional circumstances. There are many well documented cases of small flowers and shoots embedded in amber (e.g. Fig. 10.17b) sometimes together with small insects. Lignified wood still retains its structure, and charcoalification can occasionally preserve the three-dimensional aspect of plant organs. There are many conifer shoots preserved in this way in rocks as early as the Pennsylvanian age onwards, but more importantly there are also the exquisitely preserved charcoalified Cretaceous fossil flowers found in Portugal and Sweden (Fig. 10.8). Such specimens can be prepared for sectioning by infiltrating and embedding them in glycol methacrylate. Although this technique can produce excellent results it can take a long time. The infiltrating can take up to 24 hours and then takes about 4 days to harden into a strong, hard, semitransparent plastic. Large specimens can be embedded in this way and then cut into pieces before sectioning.

Prepared casts

Fossils preserved as impressions in ironstone may be preserved three-dimensionally, rather than as simple adpressions. Of course, they can be studied directly, in a similar way to adpressions, but sometimes it may be possible to gain a greater understanding of the plant remains by making a cast to produce a solid replica of the original fossil. In order to form a perfect replica it

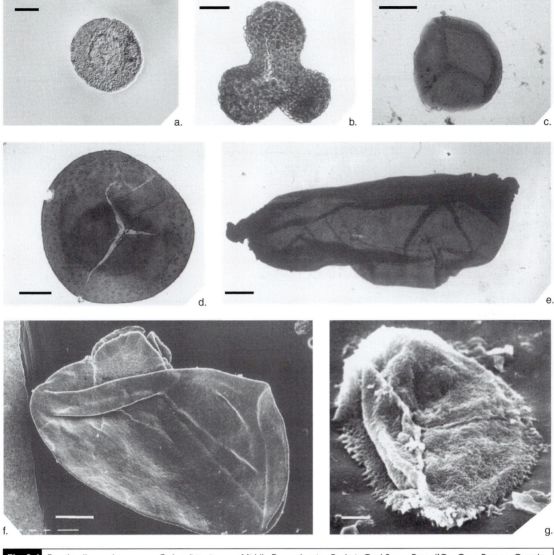

Fig. 3.4 Fossil pollen and spores. a. *Guthoerlisporites* sp., Middle Pennsylvanian Backpit Coal Seam, Bras d'Or, Cape Breton, Canada. Pollen from an extra-basinal plant, possibly a peltasperm. Scale bar = 10 µm. b. *Trilobosporites* sp is probably a fern spore, from the Lower Cretaceous Series at Hastings, southern England, with a characteristic three-lobed outline, and a densely tuberculate spore wall. Scale bar = 10 µm. c. *Cyclogranisporites* sp. is a spore from the Devonian of the Whitney Bore Hole in Oxfordshire, England. Scale bar = 10 µm. d. *Tuberculatisporites brevispiculus* (Loose) Potonié & Kremp is a megaspore from a Middle Pennsylvanian sigillarian cone from Yorkshire, England, with smooth contact faces and a tuberculate spore wall. Scale bar = 0.2 mm. e. *Cystosporites giganteus* (Zerndt) Schopf is a large megaspore from a Carboniferous *Lepidocarpon*-type cone. One of the aborted spores of the original tetrad is visible at the top right. Scale bar = 0.2 mm long. f. *Lagenoisporites* megaspore extracted from a Middle Pennsylvanian *Flemingites* cone from Mazon Creek, Illinois, USA (NMW Specimen 87.G12.1). Scale bar = 0.2 mm. g. *Lycospora* microspore extracted from a *Lepidostrobus* cone from the Lower Pennsylvanian Series of Lancashire, England (Manchester Museum Specimen LL5673). Viewed under SEM. Scale bar = 5 µm. Photos by B. A. Thomas.

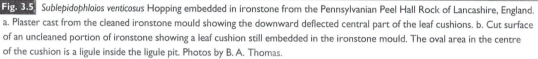

Fig. 3.5 *Sublepidophloios venticosus* Hopping embedded in ironstone from the Pennsylvanian Peel Hall Rock of Lancashire, England. a. Plaster cast from the cleaned ironstone mould showing the downward deflected central part of the leaf cushions. b. Cut surface of an uncleaned portion of ironstone showing a leaf cushion still embedded in the ironstone mould. The oval area in the centre of the cushion is a ligule inside the ligule pit. Photos by B. A. Thomas.

may be necessary to remove the remains of minerals that are still present in the mould before filling it with material to make the cast (Fig. 3.5). Plaster of Paris can be used, but low-viscosity silicone rubber is the most appropriate material. One of the best substances for making such a cast is dental impression material such as 'silflo®' which can give almost perfect surface details that can be studied by scanning electron microscope. However, caution should be used when 'cleaning' the mould in order to make a cast. Sometimes portions of the outer tissues are preserved in limonite or pyrite, and such remains can sometimes reveal anatomical detail by cutting and polishing, or thin sectioning.

Specimens that have been mineralised can reveal very fine surface detail of the fossil, and even the individual cells that can be studied by scanning electron microscopy. Sometimes mineralisations of fructifications still contain spores, which can be extracted and studied microscopically.

Sectioning anatomically preserved fossils

Originally, anatomically preserved fossils were studied using ground thin sections, similar to those used for the study of rock petrology (in fact, this technique was first used in the early nineteenth century as a palaeobotanical tool, before thin sections were used to study rock petrology). Where the cell walls are still retained as coalified layers, however, the 'peel' method may produce better results (Fig. 3.6). A flat surface of the fossil is polished and then briefly etched in acid to remove a thin layer of the mineral. The projecting coalified cell walls are flooded with acetone and a thin sheet of acetate laid over the surface. When dry, the acetate embedded cell walls can then be peeled away from the fossil. The acetate film thus has preserved in it a thin section through the fossil, which can be examined under the microscope, revealing remarkable detail of the anatomy of the plant tissue. By this means plant organs can be 'serially sectioned' to provide a detailed study of their internal anatomy.

There are many examples of fossil plants that have been preserved through impregnation with silicon-rich water, usually as a result of volcanic or geothermal activity. In such specimens the replacement may be complete so that the only way of studying them is by direct observation of polished faces or by preparing thin sections.

Pyritised and limonited specimens require techniques for both preservation and examination.

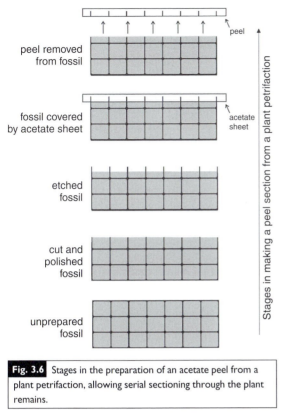

peel removed from fossil

fossil covered by acetate sheet

etched fossil

cut and polished fossil

unprepared fossil

acetate sheet

peel

Stages in making a peel section from a plant petrifaction

Fig. 3.6 Stages in the preparation of an acetate peel from a plant petrifaction, allowing serial sectioning through the plant remains.

The problem is that pyritised specimens (e.g. Fig. 3.7) can deteriorate and possibly completely decompose through oxidation. Coating with lacquer or resin is ineffective in the long term because they are pervious to both moisture and oxygen. Specimens can be kept submerged in oil with no deterioration. Some techniques for the neutralisation and removal of products of decay from pyritised specimens are potentially toxic. Immersion in 2–5% ethanolamine thioglycollate in 95% industrial methylated spirits will neutralise and dissolve ferrothioglycollate. When the reaction ceases, the specimen can be washed in clean alcohol and allowed to dry.

Pyrite permineralisations are opaque so the simplest technique is to produce surfaces for direct observation by polishing with fine abrasives, followed by etching with nitric acid to improve contrast. Peel preparations can be made by etching the polished surface with concentrated nitric acid, neutralising and then pouring onto the surface a saturated solution of polyvinyl chloride dissolved in ethyl methyl

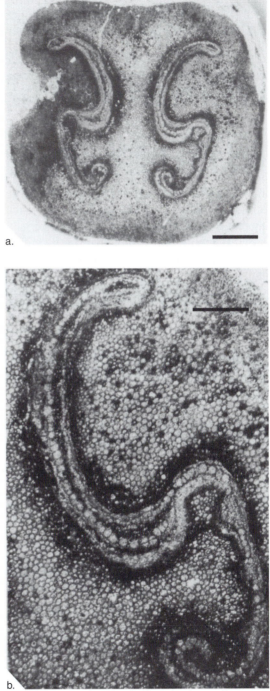

a.

b.

Fig. 3.7 Fern rachis from the Eocene London Clay deposits, on the foreshore of the Isle of Sheppey, England. a. Transverse section showing the pyritised vascular strand. Scale bar = 1 mm. b. Close-up showing cell detail. Scale bar = 0.05 mm. Photos by M. E. Collinson.

ketone. When hardened the peel can be pulled off and demineralised with concentrated nitric acid. Stein *et al.* (1982) have described other techniques for preparing pyrite and demineralised limonite sections.

Reconstructing whole fossil plants

Reconstructing the whole plant is one of the most important aspects of studying plant fossils for many palaeobotanists. It is particularly important for phylogenetic studies, because techniques such as cladistic analysis only make any real sense if they are done on whole organisms. It is important to remember, however, that virtually all published reconstructions of whole plants, including those figured in the present book, are to some extent hypothetical. The palaeobotanist has built up a model of what the plant probably looked like from a range of different types of evidence, some of which may be better than others; a number of proposed reconstructions have been shown to be incorrect. For instance, many medullosalean pteridosperms (discussed in Chapter 8) used to be reconstructed with fronds that had ovules attached to the end of their pinnae, based on a few fossils that appeared to show this. However, it now looks as though these were just examples of ovules that had accidentally fallen onto a piece of pinna as it lay in the sediment; the ovules are now thought to have been borne on quite separate branching structures attached to the distal part of the tree's trunk.

It is important to be aware of what evidence is used to suggest such connections between different organs. The best evidence is obviously where two parts of the plant are found still in organic connection. However, as already pointed out, care must be taken in interpreting such evidence in adpression fossils. Anatomically preserved fossils can be more reliable in indicating such connections, although petrifactions tend to be smaller than adpressions and the chances of finding two plant organs in connection are correspondingly rarer.

Another useful indication is the similarity in anatomical features shown by two different plant organs. One of the most famous examples of this was the discovery in the early twentieth century that fronds borne by *Lyginopteris* stems, and ovules called *Lagenostoma*, both had very similar glandular hairs on their surface. This was used to propose a new group of plants called the pteridosperms (discussed in Chapter 8). The difficulty here is that such comparisons can usually only be made between anatomically preserved fossils, and/or compressions where cuticles are still preserved. Mutual association of the plant organs can be an important corroborating factor, because there would be little sense in linking two organs if they are never found together, regardless of how similar they may seem anatomically.

Finding two plant organs regularly associated together in different macrofloras can suggest that they were from the same plant, but on its own such evidence has to be used with care unless confirmed by anatomical similarity or examples of physical attachment. It is quite possible that two plant species were closely tied to the same physical habitat, and so their remains are regularly found in close association as fossils. Nevertheless, association can be a useful guide. For instance, large trigonocarpid ovules similar to those shown in Fig. 8.11b, c have been found closely associated with fronds of *Alethopteris pseudograndinioides* var. *subzeilleri*. This is supported by the fact that similar ovules when anatomically preserved have a proximal stalk whose anatomy is very similar to the rachises of *Alethopteris zeilleri* fronds.

Naming plant fossils

The naming of specimens is a crucial aspect of palaeobotany. Without an accurate way of recording individual or collections of plant fossils, there is no way that the subject can develop beyond simple descriptions of individual objects. It is normal to name plant fossils using a system of binomial nomenclature very similar to that used in botany – a logical approach because they are, after all, the remains of once living plants.

This is not the place to go into a detailed discussion of botanical nomenclature, as there

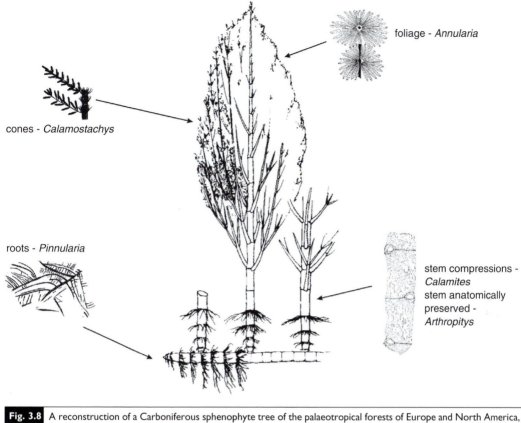

cones - *Calamostachys*

foliage - *Annularia*

roots - *Pinnularia*

stem compressions -
Calamites
stem anatomically
preserved -
Arthropitys

Fig. 3.8 A reconstruction of a Carboniferous sphenophyte tree of the palaeotropical forests of Europe and North America, showing how different parts of the plant are assigned their own fossil taxonomic names. Also, in the case of the stems, different names can be used for adpression and petrifaction fossils.

are several clear accounts of this subject (listed at the end of the chapter). Briefly, however, a plant is usually given a two-part name (binomial) consisting of the genus name and species name, e.g. the royal fern is called *Osmunda regalis*, where *Osmunda* is the genus and *regalis* is the species. It is usual to follow this by the name of the person who first named the species, in this case Linnaeus (often abbreviated as 'L.'). A species can then be assigned to a hierarchical set of higher taxa, in part reflecting the phylogenetic relationships with other taxa. In increasing rank above genus these are: family (name usually ending in -aceae), order (name usually ending in -ales), class (name usually ending in -opsida) and division or phylum (name usually ending in -ophyta). There are also intermediate ranks between these, but they rarely impinge on palaeobotany.

However, there is one important point where palaeobotanical nomenclature differs from botanical nomenclature and this can sometimes cause confusion. Plant fossils can rarely be named in the same way as living plants, because most are only fragmentary remains of the original plant and never show the full range of characters that are available in a living plant. DNA, for instance, can only be recovered in a few cases such as from plants entombed in amber, and only a small proportion of fossils have preserved cellular tissue showing good anatomical detail.

Palaeobotanists have, therefore, developed a system of nomenclature whereby isolated parts of the plant are named using separate sets of taxa, known as fossil taxa (e.g. Fig. 3.8). Different parts of the plant are assigned to different fossil genera and within each fossil genus, fossil species are based on characters of the relevant organ. Palaeobotanists usually try to make the fossil species as near to a whole-plant ('natural') species as they can, but a complete correspondence is rarely possible. The rooting structures, for instance, are virtually indistinguishable in most of the arborescent lycophytes and tend to be assigned to a single fossil species, *Stigmaria ficoides*. Fossil species of stem are based mainly on the more variable characters of the leaf-cushions. Many different stem fossil species have been recognised but some may merely represent variation within a 'natural' species or even variation between different parts of the plant.

Different modes of preservation can also give rise to the use of different fossil genera and species names for what might be parts of the same plant. As we have shown, each type of preservation can yield different types of information, so it can be difficult to be sure that one is dealing with the remains of the same original taxon. A petrifaction can provide abundant information about the detailed cell structure of the plant organ, but it can be difficult to interpret what the plant looked like as a whole. Similarly, an adpression can clearly show the shape of the organ, but may yield little of its anatomical structure. An example of this approach can be seen in naming the ovules of medullosalean pteridosperms, which are called *Pachytesta* when anatomically preserved in coal balls, and *Trigonocarpus* when preserved as adpressions or casts (for examples of such fossils, see Fig. 8.11).

There are two principal ranks of fossil taxa: species and genera. In theory, there can also be fossil families and higher ranked taxa, but these are rarely used in practice. Instead, most fossil genera are assigned to biological taxa based on whole organisms. In some cases, a fossil genus may be assignable to an order of plants, but not to a particular family; in which case it should be referred to as a satellite genus within that order.

As with biological species, a fossil species is formally tied to a particular specimen, known as its holotype, which is designated when the taxon is first formally described. If at some point that fossil species is revised and split into two distinct taxa, the original name stays with whichever of the new taxa contains the original holotype.

Some of the taxa described during the nineteenth and early twentieth centuries were based on several types now known as syntypes. It is the role of the modern-day palaeobotanist when any such taxon is revised to select one of these syntypes as a lectotype, which then effectively fulfils the role of a holotype. If the holotype or syntypes have been lost, it is in principle possible to regard the illustrations published when the species was first described as the type. However, to avoid ambiguities in the interpretation of the species it may be necessary to designate another specimen that the original author did not see, which is referred to as a neotype.

Mention must briefly be made of a particular category of fossil taxa called morphotaxa. These are defined purely on the morphology of the particular fossil to which they refer, and can only be used for fossils of the same plant part and preservation as the type specimen on which the morphotaxon was originally based. Morphotaxa are mainly of use in the study of dispersed pollen and spores, because of the difficulties of assigning some of them to particular groups of plants. However, there are a small number of morphogenera that have been developed for use with macrofossils, especially of fern-like fronds. These names mostly originated from the pioneering work of Adolphe Brongniart in the early nineteenth century (see Chapter 2) and proved useful because of the difficulty in some cases of distinguishing fern fronds from the compound leaves of some seed-plants ('pteridosperms'). For instance, *Pecopteris* is defined exclusively on the linguaeform (tongue-shaped) pinnules that are broadly attached to the rachis, and the broadly arching veins. If a fossil is assigned to a morphogenus such as *Pecopteris*, no implication is being made as to its systematic position. This is in contrast to fossil genera, where there is usually some implication being made about its systematic position: for instance, if a stem is assigned to the fossil genus *Lepidodendron*, it is implied that it belongs to the lycophytes, and probably to the Lepidocarpales. In macrofossils, morphotaxa should be seen as essentially 'holding bays' for poorly understood fossils whose systematic position is unknown. When we have learnt more about the plant that formed the fossil (e.g. reproductive structures are discovered) it is normal for the species to be transferred to a fossil genus that is more systematically meaningful. As will be discussed in Chapter 7, for instance, many of the species originally placed in *Pecopteris* have been subsequently transferred to fossil genera whose diagnoses include features of synangia, such as *Acitheca* and *Cyathocarpus*.

The main drawback of using fossil taxa is that it significantly inflates the species list for a given assemblage, and can give a misleading impression of the extent of the original biodiversity. However, the processes of transport and fossilisation that the plant fragments have been subjected to before preservation has already distorted this information so dramatically that the nomenclature issue is not really the major problem here. The greater accuracy that using fossil taxa provides to the palaeobotanist for recording plant fossils far outweighs the problem of artificially inflated biodiversity.

Another problem encountered in using fossil taxa is how to name the whole reconstructed plant. Small, herbaceous plants can sometimes be reconstructed as whole organisms (although still usually lacking many anatomical and cytological details) that can in principle be assigned a taxonomic name. It might, for instance, be possible to reconstruct a reasonably whole organism that bore a particular type (fossil species) of *Sphenophyllum* leafy shoots (see Chapter 6). However, there may have been other biological species that bore foliage that looked very similar, and which we might assign to the same fossil species of *Sphenophyllum*, but had quite a different type of spore-bearing cone. Some palaeobotanists have argued that the whole plant should be named based on the fossil taxa of the reproductive organs, but this approach is not always the best option; it has been shown, for instance, that the Carboniferous ovules *Pachytesta*/*Trigonocarpus* were borne on a range of different medullosalean plants and so would be a poor basis for naming a whole plant.

There are even greater problems with the larger tree-sized plants whose remains dominate most macrofloras. Many reconstructions have been proposed for such trees and many are illustrated in this book, but they have been synthesised from various pieces of evidence, whose quality is highly variable (see previous section). A specimen may show that a particular fossil species of cone was attached to a fossil species of stem, and another specimen show that the same fossil species of stem was attached to a particular fossil species of rooting structure. But we have no indisputable evidence that this fossil species of stem was only produced by one 'natural' species of whole plant. We have only a very few examples where we can unequivocally reconstruct all of the parts of what was originally an organic, whole-plant species of tree.

The validity of whole plant reconstructions is not completely undermined by such problems, but they should be seen for what they are: working hypotheses based on the best available data and as such do not merit formal taxonomic names. It is instead better to give these reconstructions informal names that are usually derived from the formal name of the best-known organ, such as the '*Calamites* tree' shown in Fig. 3.8, which is based on the name of the stem.

Phylogenetic analysis

Many botanists and palaeobotanists concerned with elucidating the evolutionary history of plants today use a technique known as cladistic analysis. This takes a data matrix that has the presence or absence of various morphological and anatomical characters for the different taxa being considered, and tries to find the simplest arrangement of the taxa based on the appearances and disappearances of those characters. To be meaningful, it has to be based on whole reconstructed organisms that, as we suggested earlier, are really hypothetical constructs rather than actual objects. Further information of cladistic approaches to plant evolution can be found in specialised texts, such as that by Forey *et al.* (1998).

Biostratigraphy and palaeobiogeography

Having named the fossils it is possible to compare the distribution of the taxa in time and

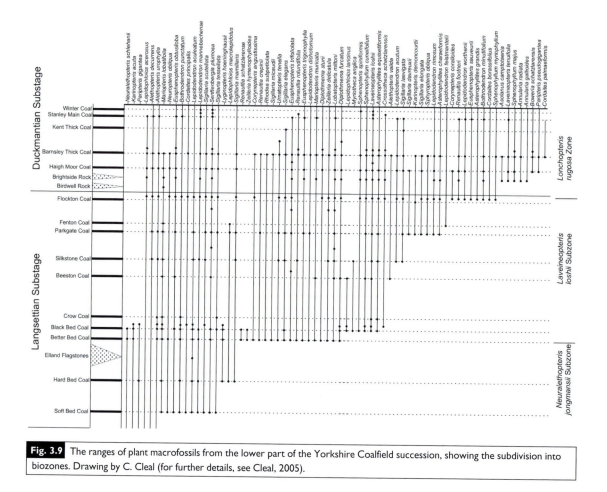

Fig. 3.9 The ranges of plant macrofossils from the lower part of the Yorkshire Coalfield succession, showing the subdivision into biozones. Drawing by C. Cleal (for further details, see Cleal, 2005).

space. The study of the temporal changes in fossil distribution as observed through the stratigraphical column is known as biostratigraphy. The biostratigraphical procedures used in palaeobotany do not differ in any way from those used in other groups of fossils, and are well-covered in standard stratigraphical textbooks. In essence, the aim is to determine the range of the taxa through the stratigraphical sequence being studied, and then to use this to divide the succession up into intervals known as (bio)zones (e.g. Fig. 3.9). If similar sequences of zones can be recognised in different stratigraphical successions, this can be used as an indication of chronostratigraphical (i.e. time-based) correlation of the successions. This might be thought to be the sole role of biostratigraphy, but that would ignore the fact that the zones also reflect the change of vegetation with time, which can provide

important evidence of both evolutionary and habitat change.

Biostratigraphical investigations on macrofloras have tended to be fairly conventional in drawing up the stratigraphical ranges of fossil taxa (usually species), recognising patterns (biozones) in their distributions, and comparing the patterns between different stratigraphical successions. More refined, statistical methods have been developed for biostratigraphy, such as graphical correlation and Unitary Associations analysis but although these have been used in palynological studies, we know of no examples of these being used for macrofloras.

Palaeobiogeography is concerned with identifying geographical patterns in the distribution of fossils. Just as biostratigraphy can be used to help establish time-correlations between stratigraphical sequences, palaeobiogeography has been used as a guide to establishing palaeogeographical models. As indicated earlier, Wegener's continental drift ideas were partly based on evidence of plant palaeobiogeography. However, palaeobiogeography can also give evidence of the biogeography of past vegetation, which in turn can help us understand past environments and climate, as well as plant evolution.

The traditional approach has been to divide up a geographical area into a series of palaeofloristic units known as phytochoria – areas that have comparable macrofloras. As with the biogeography of living plants, this was done using a nested hierarchy of units, in descending order of rank: palaeokingdom, palaeoarea, palaeoprovince and palaeodistrict. These are similar to the terms used for the biogeography of modern vegetation, but with the suffix palaeo- added to emphasise that they are conceptually different because of the distortion introduced by the fossil record; fossil phytochoria are almost exclusively based on lowland vegetation, whereas modern phytochoria are based on all vegetation. It is also common to refer to palaeolatitudes, meaning the latitude that a flora was at, at the time that it was growing; as we show in Chapter 11, this can often be quite different to the latitude where the locality is today.

There is no universally accepted rule as to how palaeofloristic units are assigned to a particular rank. We suggest that if it is defined by its content at a largely supra-generic level, it should be regarded as a palaeokingdom; if it is defined largely by differences of genera then it is a palaeoarea; if by a large number of species then it is a palaeoprovince; and if by just a few species then a palaeodistrict. This is not meant to be a rigid definition, but merely reflects the way that we have approached this problem. Examples of palaeogeographical maps showing the distribution of phytochoria are illustrated in Chapter 11.

Recently, there have been attempts to fit fossil floras into 'biomes'. Modern-day biomes are defined as distinctive regional groups of animals and plants adapted to certain broad physical conditions, e.g. boreal taiga/forests, mangroves. The problem with trying to use this approach for the plant fossil record is similar to the problem of using floristic units – the fossil record is just too incomplete to allow a proper understanding of the whole biota. There is also the problem that our understanding of the physical conditions to which the biotas were adapted is at least partly derived from observations on those biotas, so there is clearly the potential for circular reasoning.

The tradition in the first half of the twentieth century was to use species lists from different localities to interpret palaeobiogeographical patterns. The problem with this approach is that there is no easy graphical way to present the geographical distribution of many species. This is in contrast to the range charts that are used to identify patterns in the stratigraphical distribution of fossils. There is, therefore, an increasing tendency to use mathematical methods to recognise geographical patterns. These fall broadly into two groups. There are hierarchical methods, notably cluster analysis, that produce dendrograms showing how closely the macrofloras of different localities (or areas) correspond. This method compares all possible pairs of macrofloras in the available data, and for each pair calculates a similarity coefficient that gives a measure of the relative number of species they have and have not in common. Various coefficients are available for this analysis, but Raup-Crick coefficients have proved especially useful for the type of presence–absence data that are usually available. The resulting matrix of similarity coefficients is then analysed to determine how the localities group together, and the pattern represented as a dendrogram (e.g. Fig. 3.10). The results of a cluster analysis are usually easy to interpret, but it only resolves the pattern in one dimension; if there are two or more factors affecting plant distribution (e.g. latitude, longitude, substrate quality, etc.) then the analysis may not be able to interpret all the variables.

Ordination analysis arranges the macrofloras in two or more dimensional graphs; the nearer they plot together on the graph, the more similar they are. Various ordination methods are available. The conceptually simplest is Detrended Correspondence Analysis (DCA), which attempts to sort the localities into a two- or more dimensional space, with each axis assumed to represent an ecological or biogeographical gradient (Fig. 3.10). Principle Coordinates Analysis and

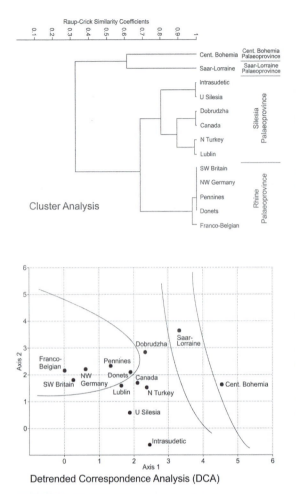

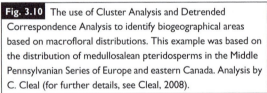

Detrended Correspondence Analysis (DCA)

Fig. 3.10 The use of Cluster Analysis and Detrended Correspondence Analysis to identify biogeographical areas based on macrofloral distributions. This example was based on the distribution of medullosalean pteridosperms in the Middle Pennsylvanian Series of Europe and eastern Canada. Analysis by C. Cleal (for further details, see Cleal, 2008).

Non-metric Multidimensional Scaling use different methods to ordinate the data based on the same types of similarity coefficients as used in cluster analysis. Principle Components Analysis tries to find linear patterns in the data based on a mathematical technique known as eigenvalue analysis.

There is no 'right' or 'wrong' technique to use for analysing palaeobiogeographical data and it is often worth trying different analyses. Always remember that one is not necessarily better than the other, they are just showing different pictures of the structure within the dataset being analysed.

Further information on all of these techniques can be found in Hammer & Harper (2006).

Some modern plant biogeographers have adopted cladistic methods of analysis, essentially similar to those used in phylogenetic analysis. Like cluster analysis, the output is a dendrogram; but, unlike cluster analysis, the groups are not based on general similarities between the floras, but on the progressive appearance and/or disappearance of individual taxa. The few attempts that have been made to use this approach with macrofloral palaeobiogeography have not given significantly different results to those obtained from cluster analysis.

Curation

Basic curation of any plant fossil specimen is critical if it is to retain the full range of evidence that it can provide. Firstly, there is the conservation of the specimen, i.e. ensuring it is kept in the correct conditions so that it degrades as little as possible. Secondly, there is the documentation – ensuring the specimen is correctly labelled and numbered; without information as to exactly where it was found, most plant fossils are virtually worthless. Type, figured and referenced specimens should be numbered and registered in recognised collections, and if at all possible in a national museum. Curation is an extensive subject in itself, and for further details the reader is guided to the *Guidelines for the curation of geological materials* published by The Geological Society, London.

Site conservation

For most people, palaeobotany is very much a field-based activity and without continued collecting the amount of new data that becomes available will be limited. It is important, therefore, for palaeobotanists to be aware of the various pressures on the physical nature of plant fossil sites.

Notwithstanding the fact that localities are very rarely worked out, there is the real danger that some may be lost through development or landfill. A few sites such as the Fossil Grove in

a.

b.

Fig. 3.11 a. Joggins World Heritage Site, Nova Scotia, Canada. General view of some of the cliffs with *in situ* tree stumps of Late Carboniferous age. Photo by C. Cleal, taken during an excursion with the 2003 North American Paleontological Convention. b. Lesbos Geopark, Greece. A standing conifer tree trunk, 4.4 m tall, and identified as *Taxodioxylon gypsaceum* (Göppert) Kräusel. It is covered by pyroclastic material (mudflow) due to a volcanic explosion during early Neogene times (*c.* 18 Ma). Photo by N. Zouros. Reproduced with permission.

Glasgow and the Petrified Forest in Arizona have been protected in different ways for many years, but the vast majority have not been protected. Many localities of palaeobotanical interest such as working mines or opencasts can of course never be conserved. For further details, see Thomas (2005) and the various papers in Bureck & Prosser (2008).

In recent years, there has been an increasing desire among geologists to protect the best of the sites and some countries have initiated schemes for such protection. In Britain, the Geological Conservation Review was initiated in 1977 by the then Nature Conservancy Council to assess, document and publish accounts of the most important geological sites. These GCR selected sites have been included in the lists of Sites of Special Scientific Interest (SSSIs) notified under the Wildlife and Countryside Act 1981. A series of volumes describing the GCR sites has been variously sponsored by or published by the Joint Nature Conservation Committee, and two of these covered palaeobotanical sites (written wholly or in part by the present authors).

The situation is rather different in the United States. There are many Federal and State regulations and laws protecting sites and governing the collection of natural history objects. State owned land is managed by the Department of the Interior which can control collecting through the issue of permits. Wolberg and Reinhard (1997) give a useful summary of the current situation.

There have also been attempts to give international designations to geological sites, to help protect them. For example, a number have been given World Heritage status, one of which has significant palaeobotanical interest – Joggins section in Nova Scotia, Canada (Fig. 3.11a). To further improve the scientific basis for the selection of sites for World Heritage status, the International Union of Geological Sciences has initiated GEOSITES – a project to compile national and regional inventories of sites from which it should be possible to identify those of global significance. Regional working groups have been established throughout the world to develop these inventories. In the UK, for instance, the British Institute for Geological Conservation is undertaking this part of the work. Another international designation is as a UNESCO Geopark, which aims to combine conservation of the sites with community involvement and sustainable development through geotourism. The most notable example of a geopark with palaeobotanical interest is the Neogene fossil forest at Lesbos, Greece (Fig. 3.11b).

Recommended reading

Brunton et al. (1985), Cleal (2005, 2008), Cornish & Doyle (1984), Crabb (2001), Cridland & Williams (1966), Hammer & Harper (2006), Jansonius & McGregor (1996), Jennings (1972), Jones & Rowe (1999), Kerp (1991), Kerp & Barthel (1993), Krings (2000), Lacey (1963), Poinar et al. (1993), Robinson & Miller (1975), Stein et al. (1982), Thomas et al. (2004), Traverse (2007), Wolberg & Reinhard (1997).

Chapter 4

Early land plants

The first organisms to make any real impact on the land surface were plants. Evidence from fossil soils (we are using the term here in a general sense for unconsolidated subaerial sediment lying on bedrock) suggests that some bacteria and protoctist algae were able to obtain a foothold on the land surface before plants, but they were small and had very limited impact on the physical environment. Until there was plant-cover to provide food and shelter, animals were essentially aquatic organisms. The rooting and anchoring structures of plants helped stabilise terrestrial sedimentary environments and increased the development of soils. Although aquatic photosynthetic algae had been having an important impact on the composition of the atmosphere since Precambrian times (Fig. 4.1), especially its carbon dioxide content, the appearance of larger land-plants significantly enhanced this process. The first invasion of the land by upright and vascularised plants was one of the key events in the evolution of life, as it was a necessary precursor for the development of terrestrial life and habitats as we know them today.

Alternating generations

Living plants show a vast array of different modes of reproduction, from the simpler strategies of ferns and other spore-producing plants, where water remains a crucial part of the process of fertilisation, to the specialised strategies shown by many flowering plants that rely on animals or wind for pollination prior to pollen tube growth that leads to fertilisation. However, underlying all plant reproduction is the same fundamental pattern of alternating sexual and asexual phases that make up their life cycle (Fig. 4.2).

The simplest form of alternation of generations is when the recognisable plant (the sporophyte) produces and releases a single type of spore as its primary means of dispersal. The spores are produced in a sporangium by a process of cell-division known as meiosis, which results in clusters of four spores (tetrads) with each one having just a single set of chromosomes (i.e. haploid). Spores formed in isodiametric tetrads have three faces where they were in contact

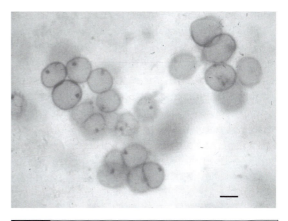

Fig. 4.1 Very early photosynthetic unicellular eukaryotic (nucleate) organisms, *Glenobotridium*, from the Precambrian Bitter Springs Formation of central Australia, about 900 million years old. Scale bar = 10 μm. Each cell contains an eccentrically located body that resembles a starch-enclosed pyrenoid. Photo provided by W. G. Chaloner.

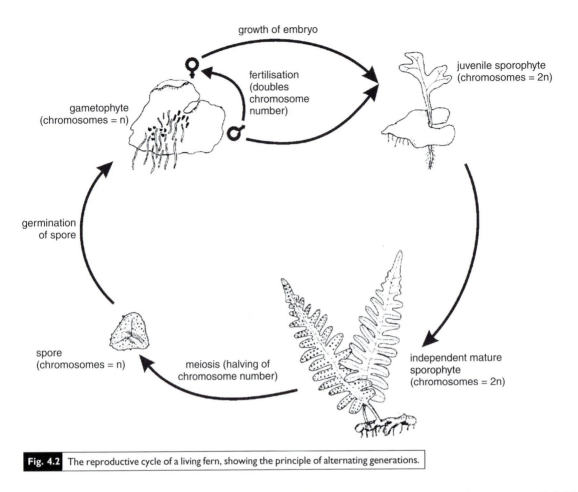

Fig. 4.2 The reproductive cycle of a living fern, showing the principle of alternating generations.

with the three other members of the tetrad. The ridge, or line, between the three contact faces is called the triradiate mark and the spore wall usually splits along these lines when it germinates. If a spore finds a suitable place to germinate, it grows to form the plant's sexual gametophyte phase, the prothallus. The prothallus consists of haploid cells and bears both types of sexual organs known as archegonia (female) and antheridia (male). The sex organs produce male and female gametes analogous to sperm and eggs in animals. The male gametes need surface water in which to swim when they are released from the antheridia. The archegonia release chemicals that attract the swimming sperm towards them and into their open necks. One sperm will then fertilise the female gamete resulting in a zygote. This is a diploid cell, in which the chromosomes are once again in pairs. The zygote then undergoes cell division to form

the embryo that will grow into a new diploid sporophyte starting the cycle all over again.

Plants producing only one kind of spore are said to be homosporous. The earliest land plants were homosporous and we presume that their haploid spores grew into thalloid or cylindrical bisexual prothalli as described above. Later other plants developed that produced two kinds of spores and are said to be heterosporous. In living heterosporous plants, such as *Selaginella*, *Isoetes* and the water ferns, the larger megaspores open along their triradiate marks to reveal an archegonium-bearing prothallus which remains within the spore. The much smaller microspores produce a single antheridium which is also retained within the spore wall. These unisexual gametophytes retained within their spores are said to be endosporic while the bisexual prothalli that grow into independent plants are exosporic.

The gametophyte needs surface water for the motile male gametes to be able to swim to the archegonia, so they need moist conditions for growth and reproduction. The sporophyte, in growing from a fertilised ovum in one of the gametophyte's archegonia, is initially reliant on the gametophyte, but as a mature plant can survive drier conditions. Transient wet or moist areas may allow the gametophyte to grow and complete reproduction, and the sporophyte to grow before drier conditions prevail. Once they have established themselves sporophytes can continue to grow for many years while vegetative growth may permit them to spread asexually over large areas. However, in order to establish new colonies the sporophytes liberate spores when the air is dry allowing them to be carried by the wind to new habitats that may be suitable for gametophyte growth.

The progressive change from essentially aquatic organisms (algae) to organisms adapted to life on dry land can be seen as a change in emphasis given to the two phases in their life cycle. In most aquatic algae, the gametophyte is the dominant phase, with the sporophyte represented only as an ephemeral organism. In bryophytes, the gametophyte phase is also dominant, being the leafy or thalloid plants that we normally associate with mosses and liverworts; the sporophyte is merely the spore-bearing capsule, sometimes borne on an axis, that grows from the fertilised embryo in a gametophyte archegonium. Except for a small amount of photosynthetic activity in some bryophyte sporophytes, they are totally dependent on the gametophyte for nutrition and cannot live independently of it. This tends to mean that most bryophytes can only live in relatively damp conditions.

In land plants, however, we see the sporophyte, with its better adaptation to drier conditions, becoming more important in the life cycle. The earliest land plants had sporophytes that became as large as the gametophytes, and may have had as long a life. The trend rapidly continued, and soon the sporophyte came to dominate the life cycle. In living pteridophyte plants, such as ferns, what we normally see is the sporophyte, and the gametophyte is small and usually ephemeral. As will be discussed in Chapter 8, this process has been carried further in seed-plants, where gametophytes are never independent organisms.

Adapting to life on land

Conditions on the land were very different from the aquatic environment in which the ancestral plants lived. Some of these plants must have been preadapted in order to survive in these new conditions. They had to support themselves, avoid drying out, and be able to transport water internally. In early plants, only the sporophyte became fully adapted to living in drier conditions; the gametophyte, as we have seen, continued to need moisture for fertilisation. Most 'lower' (i.e. non-seed) plants are thus still essentially amphibious, being adapted to spending their life out of water, but unable to reproduce without it. This break from needing surface water for fertilisation was not properly achieved until the evolution of the seed (see Chapter 8).

Very small plants can support themselves in an upright position by maintaining a sufficiently high internal water pressure ('turgor'), but for larger plants this is not enough. Some modern-day mosses strengthen the outer part of the stem with tissue known as sterome and some early land plants seem to have adopted the same strategy. However, the more successful anatomical adaptation was the development of a central vascular strand or stele consisting of xylem, phloem and cambium. Primary growth in a plant occurs at the tip of the stem, where one or more cells known as an apical meristem continuously divide to produce new tissue. The development of cells with lignified thickenings on the walls in the earliest land plants provided the additional rigidity necessary for support. Xylem provides both a conduit for water and mineral salts up the stem to the areas of the plant involved in photosynthesis and mechanical support for the stem. The simplest type of xylem conducting cells are tracheids whose walls have distinctive thickenings of lignin that can be helical, annular or scalariform ('ladder-like'). These are found in the earliest vascular land plants. More advanced tracheids are completely lignified except for localised holes called pits, which connect neighbouring tracheids to provide the route for the passage of water through the xylem. Tracheids can sometimes be recognised as isolated fragments in the fossil record, although care has to be taken not to confuse them with parts of certain animal fossils, such as graptolites (such mistakes have resulted in a number of erroneous records of apparently early plants in marine strata). The part of the stele involved in transporting fluids away after photosynthesis is the phloem, which consists of living thin-walled cells and has no strengthening function.

Even though water can be transported throughout plants by xylem, the amount available may not be enough and this could lead to the dehydration and collapse of the plants. Dehydration is limited by a protective outer cuticle that consists of a layer of different long-chain fatty acids and their polyhydroxy derivatives secreted by the epidermis. This both limits dehydration of the plant tissue and provides protection against attack by pathogens. However, plants cannot be hermetically sealed from the atmosphere, as their physiology requires the intake of carbon dioxide for photosynthesis and oxygen for respiration. They also need to lose some water to activate the transpiration stream, which brings minerals from the substratum. Thin leaves can absorb gases directly through the cuticle and lose water by evaporation. Larger thicker leaves have a lower surface area to volume ratio that does not allow enough simple transference of gas and water through the surface. Instead, gaseous exchange and water vapour loss are through small pores in the surface of the plant. Clearly, the plant needs to balance gaseous exchange against water loss to survive; it must be able to open the pores when it needs to exchange gases with the atmosphere and there is sufficient water for transpiration, but narrow or close them when water needs to be conserved. In nearly all plants, this is achieved by a pair of cells that surrounds the pore, and can control the size of the pore according to needs; these cells are known as guard cells. The pores and their guard cells are together known as stomata (singular stoma; see

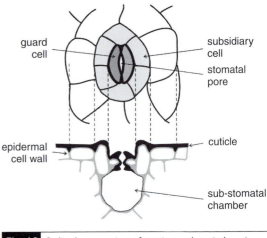

Fig. 4.3 Stylised stomata in surface view and vertical section.

Fig. 4.3). To increase the efficiency of water loss and gaseous exchange there is usually a sub-stomatal cavity that connects with an extensive system of intercellular spaces in the subepidermal cortex.

The spores produced by the sporophyte for dispersal also need some protection from dehydration as there may be a long wait between release by the parent and eventual germination. In the latter stages of their development in the sporangia, spores are covered by a protective layer of a polymerised carotenoid known as sporopollenin, which can withstand dehydration and high temperatures.

The detailed pattern of evolution of the various adaptive features that culminated in the appearance of land vascular plants remains controversial. Spores appear significantly earlier in the fossil record than macrofossils. This may be due to a lack of suitable terrestrial deposits representing the intervening time interval, but this is unlikely because many macrofossils of early land plants are allochthonous in shallow marine or marginal marine deposits. A more likely explanation is that spores adapted to dry conditions indeed evolved first, followed later by sporophytes that could also withstand dry conditions to allow effective release of the spores into the atmosphere. Then, by a progressive shift in emphasis in the life cycle of plants from the gametophyte to the sporophyte, plants as a whole became better adapted to life on dry land.

Cryptospores and the earliest land plants

Llanvirn-age rocks (early Ordovician Period, *c.* 480 million years ago) in the Czech Republic, Saudi Arabia and Oman have yielded microfossils known as cryptospores. Although the cryptospores look superficially like the spores of living vascular plants, they were dispersed as clusters of four (tetrads), two (diads) or single (monads), with the clusters in many cases being surrounded by a tight-fitting envelope of tissue. Recent work has shown that the ultrastructure of the cryptospore walls is remarkably similar to that of liverwort spores, which are also often surrounded by an envelope of tissue. Moreover, the Oman cryptospores were sometimes found in disc-shaped spore-masses, sometimes with the remains of an outer covering, similar to the sporangia of liverworts (also of other bryophytes and vascular plants). It seems likely that these cryptospores are the remains of liverwort-like plants that were showing the first adaptations to life on land – they may still have lived in water, perhaps in shallow pools, but their spores could be dispersed by air without drying-out.

It would be going too far to say that these earliest land plants were true liverworts, although it is of note that some recent molecular DNA analyses place liverworts at the very base of the phylogenetic tree of plants. It would be safer to suggest that liverworts retain more of the primitive features of these early organisms than other plants. Studies on the DNA and reproductive biology of living land plants suggest that liverworts, like the Charales (the 'stoneworts'), probably arose from aquatic green algae. The living charalean alga *Coleochaete*, which grows in various aquatic environments as a disc-shaped thallus, may well have a similar general morphology to these very early liverwort-like terrestrial plants. No fossils of *Coleochaete* or similar algae are known from Ordovician or Silurian rocks; they would almost certainly be too delicate to become fossilised. However, there is an enigmatic plant-like fossil found widely in Lower Devonian rocks of northern Europe that might offer some insights into the precursors of land plants. *Parka decipiens* (Fig. 4.4) had a disc-like thallus, up to 30 mm in diameter, consisting of two or three layers of cells. On the upper surface of the disc are numerous, small pillbox-shaped structures containing spores. The structure of the thallus and the spores is very similar to *Coleochaete*, but seems to have been better adapted to surviving drier conditions. *Parka* is found in strata with mud-cracks, suggesting that it may have grown in shallow pools that were periodically subject to drying-out. The spores were probably released when the water dried out, enabling them to be blown into other pools where they could germinate and grow into new plants. *Parka* itself is probably a descendant of the ancestral group that gave rise to the land plants.

The first vascular plants

The first hint of true vascular plants is found in early Silurian rocks (Figs. 4.5, 4.6). The cryptospores are now more consistently found singly and lacking the outer envelope of surrounding tissue (isolated Ordovician cryptospores lacking their enveloping tissue are known, but were probably the result of cryptospore-clusters

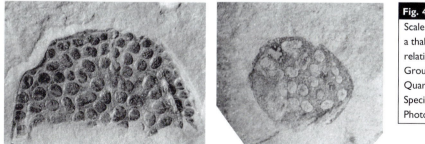

Fig. 4.4 *Parka decipiens* Fleming. Scale bars = 5 mm. Two examples of a thalloid plant, possibly a primitive relative of the bryophytes. Forfar Group (Lower Devonian), Balgavies Quarry, Forfar, Scotland (BMNH Specimens OR42665 and V.57875). Photo by A. Hemsley.

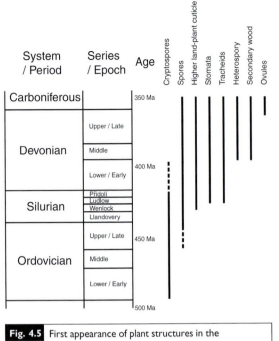

Fig. 4.5 First appearance of plant structures in the Palaeozoic Era.

Fig. 4.6 Stratigraphical ranges of main orders of early land plants through the Silurian and Devonian periods.

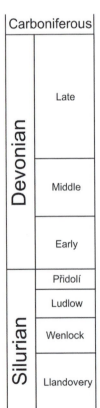

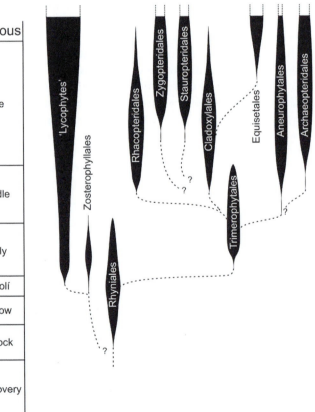

having been broken-up during preservation). The tetrad (crypto)spores, in particular, now look very similar to the spores of many living land plants. Although the details of the evolutionary processes taking place are still not fully clear (the evolutionary position of the diads, which have no modern-day counterparts, is especially puzzling) what we seem to be seeing is the progressive development of meiosis as now occurs in land plants.

Remains of the first land plants themselves can be found in early Silurian rocks. Slender branching axes of what are assumed to be land plants have been reported from the Llandovery strata, and in the Wenlock strata we have the first evidence of reproductive structures of a plant known as *Cooksonia*. This appearance of plant macrofossils probably reflects the evolution of new tissues (notably cuticle) that adapted them to surviving in drier conditions and which made them more likely to withstand fossilisation; it is notable

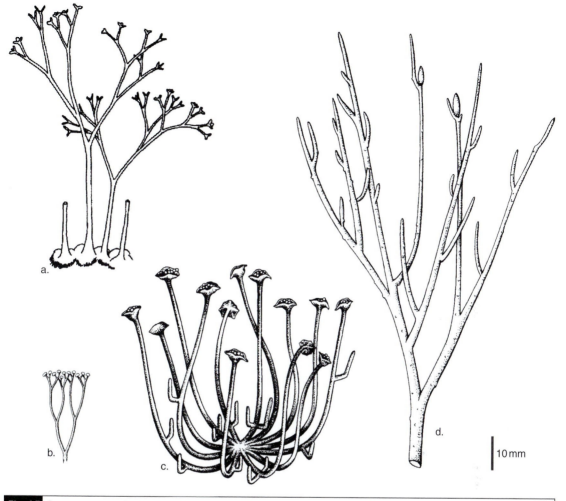

that fragments of plant cuticle also first start appearing in palynological preparations from early Silurian rocks.

It is difficult to imagine a more primitive-looking vascular plant than *Cooksonia*. It appears to have been only a few millimetres tall and consisted of smooth slender forked stems covered with cuticle and stomata. Some of the stems had a small sporangium at their tip (Figs. 4.7b, 4.8a-b). These sporangia were simple, spore-bearing sacs with no specialised structure to facilitate their opening-up (dehiscence) to release the spores. The sporangia-bearing stems presumably arose from some sort of basal structure, either a thalloid-like body or a creeping network of horizontal stems, but we have no direct evidence of this. The fossils are so simple-looking that, for many years after their first discovery, it was not certain that they were truly vascular plants. There is now unequivocal evidence that at least some were vascular plants, as the remains of tracheids have been found in Devonian specimens of *Cooksonia*. However, whether the Silurian examples also had

Box 4.1 | Identifying rhyniophytes

Rhyniophytes were morphologically very simple plants, so their fossils have relatively few characters to aid identification. This makes it both easier and more difficult to work with these fossils. Fragments of axis lacking reproductive organs are virtually impossible to assign to meaningful taxa. Axes without any form of surface ornamentation are usually referred to as *Hostinella*, while spiny axes may be called *Psilophytites* (although without evidence of reproductive structures, it is impossible to be certain that they are rhyniophytes). If the stems have terminal sporangia, then the shape of the sporangia and the pattern of branching of the stem below the sporangium are the key characters. The most widely reported genus is *Cooksonia*, which is characterised by round or somewhat flattened sporangia and dichotomously branched stems (Fig. 4.8a-b). *Steganotheca* is morphologically similar, but has more elongate sporangia with a distal thickening (Fig. 4.8c). Knowledge of the spore-content of the sporangia can give even better insights into the affinities of such fossils but this requires specialist laboratory techniques.

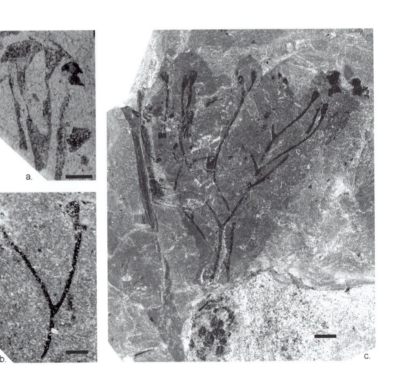

Fig. 4.8 a. *Cooksonia pertoni* Lang from its type locality, clearly showing the characteristic flattened sporangia. Scale bar = 2 mm. Upper Perton Formation (Přídolí Stage), Perton Lane, Herefordshire, England (NMW Specimen 77.6G.114). b. *C hemispherica* Lang, which has more rounded sporangia borne on a gradually widening stem. Scale bar = 2 mm. Milford Haven Group (Přídolí Stage), Freshwater East, Pembrokeshire, Wales (NMW Specimen 77.6G.27a). c. *Steganotheca striata* Edwards, which looks rather like *Cooksonia*, except that it was rather larger and had more elongate sporangia. Scale bar = 5 mm. Upper Roman Camp Formation (upper Ludlow Stage), Capel Horeb, near Llandovery, Wales (NMW Specimen 69.64G.32a). From Cleal & Thomas (1999).

vascular tissue remains to be proved. Different plants that have been classified as *Cooksonia* have been found to yield quite different types of spores and it seems that the genus may include species that are not that closely related – indeed some may be vascular, while others maybe were not.

Vascular *Cooksonia* plants are classified in the most primitive division of vascular plants, the Rhyniophyta (this division is discussed later in the section dealing with Rhynie Chert). However, there are other plant remains found in strata of similar age that are broadly similar in habit to *Cooksonia*, but are not known for certain to be vascular and so cannot be classified in the Rhyniophyta. Such plants are referred to as rhyniophytoids, i.e. they appear morphologically similar to rhyniophytes, but have no demonstrable vascular tissue. Many examples of rhyniophytoid plants have been found, but the best known is probably *Steganotheca* (Fig. 4.8c) from upper Silurian rocks in Wales. *Steganotheca* has many features in common with *Cooksonia*, including slender smooth forking stems with terminal sporangia, but it was a little more robust than *Cooksonia* with the largest known specimen being 45 mm long. The most significant difference between the two is that the sporangia of *Steganotheca* are more elongate and have truncated apices.

True rhyniophytes became extinct at the end of Early Devonian times, although rhyniophytoid-like plant-remains are found in Upper Devonian strata. The rhyniophytes may have been superficially insignificant in appearance, but they are of fundamental evolutionary importance being the group in which three key innovations that adapted plants for life on land first appeared: vascular tissue, cuticles and stomata.

The Rhynie Chert flora

Late Silurian and Early Devonian land plant remains have been known since research began into rocks of that age in the early 1800s. The problem was that the fossils were small and showed relatively few characters. Indeed because the preservation was often poor, it was difficult to be certain that they actually were the remains of primitive land plants and not just poorly preserved fragments of other plants such as mosses, or even algae. The debate continued through the nineteenth century and, with the limited data then available, the problem seemed intractable.

The breakthrough came with the discovery of the Rhynie Chert flora in the early twentieth century. Rhynie Chert was formed in an area of volcanic activity and the fossils within it represent an almost entire ecosystem, including plants, fungi and animals, that was preserved *in situ*. The area was periodically flooded by water rich in silica and other minerals, produced by nearby volcanic activity, and the silica and minerals infiltrated the plants and animals preserving them in exquisite detail. The Rhynie plants are unique to this locality, even at the generic level. It is nevertheless possible to relate the evidence obtained from Rhynie to the more abundant but less well-preserved adpression floras found in Silurian and Lower Devonian rocks, allowing us to make more botanically meaningful inferences about the affinities and evolution of these early plants.

Rhynia gwynne-vaughanii is the commonest species at Rhynie (Figs. 4.7d, 4.9a), and is arguably the most important palaeobotanically, providing the basis for the concept of the Rhyniophyta. The plant was relatively small, just a few centimetres tall. Its basal parts consisted of a mass of creeping rhizomes from which arose slender vertical stems, which showed both dichotomous and lateral branching. There were no leaves or spines on the stems, but there were small bulges along their length. The anatomy of the stems was exceedingly simple, with a slender central stele, surrounded by a two-layered cortex and a thin epidermis with stomata. Sporangia were positioned at the ends of the stems, although a lateral branch sometimes grew from just below a sporangium making it appear to have been laterally attached. As with earlier rhyniophytes such as *Cooksonia*, the sporangia were simple, with no specialised structure to facilitate the release of the spores.

Horneophyton (Fig. 4.7a) differs from *Rhynia* in two main features. Firstly, the basal part of the plant consisted of a small corm-like rhizome. The sporangia, although borne terminally on the stems, were also rather different. They were in effect little more than cavities in the end of the stem into which the stele extended to form a columella-like structure similar to that found in many mosses. The *Horneophyton* sporangium, unlike that in *Rhynia*, has a clear dehiscence structure in its apical part to help with the release of the spores. The affinities of *Horneophyton* are still uncertain. It is obviously a primitive vascular plant, but some have argued that it shows several features (e.g. the sporangia) connecting it with the bryophytes. Others place *Horneophyton* in a class of its own (Horneophytopsida) within the Rhyniophytina.

Various other land plants are known from Rhynie. Best-known is a plant that is superficially rather similar to *Rhynia*, and which was originally

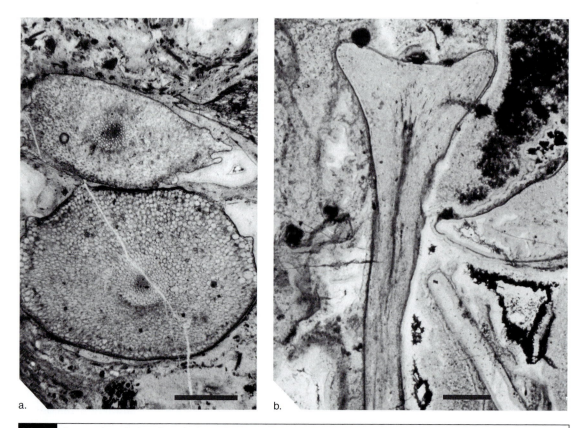

Fig. 4.9 Sections through plants from the Lower Devonian Rhynie Chert, Scotland. Scale bars = 1 mm. a. Transverse section through two stems of the sporophyte *Rhynia gwynnevaughanii* Kidston and Lang, showing a slender central vascular strand surrounded by thick cortical tissue, and irregular protuberances from the surface of the stem (NMW Specimen 21.14G.13). Photo by B. A. Thomas. b. Longitudinal section through the gametophyte *Lyonophyton rhyniensis* Remy and Remy, showing cup-shaped distal end to which antheridia were attached (specimen in the collections of the Geologisch-Paläontologisches Institut und Museum, Westfälisches Wilhelms-Universität, Münster, Germany). Photo provided by H. Kerp.

named *Rhynia major*. However, subsequent work has shown that unlike true *Rhynia* it did not have a true stele (i.e. it was rhyniophytoid and not a proper rhyniophyte) and has thus been renamed *Aglaophyton majus*. Another plant, known as *Asteroxylon*, is an early lycophyte and will be dealt with further in Chapter 5.

The nature of the gametophyte in early vascular plants was for many years debatable. Was it a small, ephemeral structure such as found in most living ferns, or a more substantial structure as seen in bryophytes? The answer came in the early 1990s when the German palaeobotanist Winfried Remy and his colleagues described several gametophytes in the Rhynie Chert. These were of a similar size to the sporophytes, with sexual organs borne on

vascularised stems, and clearly quite unlike those of any living gametophytes (Fig. 4.7c, 4.9b). So far, gametophyte equivalents have been found for three Rhynie species, including *Horneophyton*, although not yet for *Rhynia*.

Zosterophylls

Another group of primitive vascular plants appears in the fossil record, very shortly after the appearance of *Cooksonia* (Fig. 4.6). This is what we call the zosterophylls (or, formally, the Zosterophyllopsida, named after the earliest described form, *Zosterophyllum*). Many zosterophylls appear superficially similar to *Cooksonia*, having

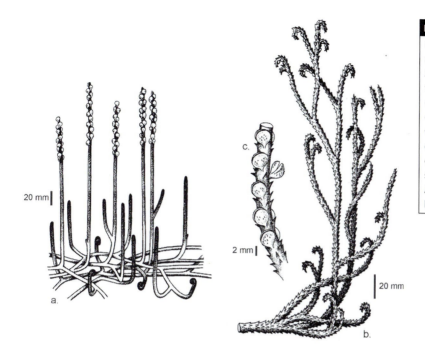

Fig. 4.10 Reconstructions of examples of Early Devonian zosterophyllaleans. a. *Zosterophyllum*, showing stems without any spines or other emergences, and terminal clusters of sporangia. b. *Sawdonia*, showing the stems with dense spines. c. Close-up of part of a fertile shoot of *Sawdonia*, showing the bivalved sporangia. Drawings by A. Townsend, based on the work of P. G. Gensel and H. N. Andrews.

Box 4.2 | Identifying zosterophylls

The zosterophyll group is best recognised in practice by the presence of stems with laterally borne sporangia, where the sporangia consist of two valves with a clear line of dehiscence. Important characters for distinguishing genera within the group are the position of attachment of the sporangia, and the surface ornamentation of the stems. For example, *Zosterophyllum* has smooth stems and sporangia arranged in clusters at the distal ends of the stems (Fig. 4.10a). *Gosslingia* also has essentially smooth stems (although they do have some small protuberances), but the sporangia are attached laterally along the length of the stems (Fig. 4.11c). *Sawdonia* is distinguished mainly by its spinose stems (Fig. 4.10b).

smooth, dichotomous stems (Fig. 4.10a). However, unlike in *Cooksonia*, there is good evidence of the basal part of the zosterophyll plant, which consisted of a prostrate mass of stems. The zosterophylls also appear to have been rather larger plants than *Cooksonia*, probably anything up to 0.5 m tall. More significantly, the zosterophyll sporangia were not borne at the ends of the stems, but were laterally attached to them (Fig. 4.10b). The sporangia were also somewhat more sophisticated than those of *Cooksonia*, consisting of two 'valves' which split apart along a definite dehiscence line to release the spores.

Zosterophyllum was first interpreted as a semi-aquatic plant and illustrated with just the terminal fertile parts of the stems protruding above the water. This now seems unlikely, as the stems were covered with stomata, so it is most likely that they were fully terrestrial plants. The stomata are of interest because they appear to have just a single, annular guard cell, unlike all other vascular plants, which have pairs of guard cells.

One variation on the zosterophyll theme is *Gosslingia breconensis* from Early Devonian macrofloras of Wales (Fig. 4.11c). This has virtually naked stems, rather like *Zosterophyllum*, and the individual sporangia are quite similar. However, instead of the sporangia being clustered near the ends of stems, they are distributed along the length of the stem.

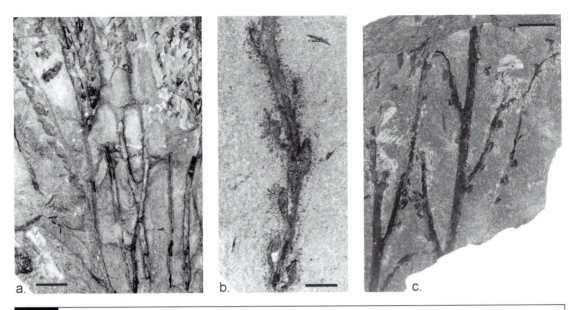

Fig. 4.11 Zosterophylls. a. *Zosterophyllum llanoveranum* Croft and Lang, showing the sporangia borne laterally on the otherwise naked axes. Scale bar = 10 mm. Lower Devonian of Llanover Quarry, near Abergavenny, Wales (BMNH Specimen V.26516a).
b. *Zosterophyllum myretonianum* Penhallow, showing a close-up of the bivalved sporangia attached laterally to the axis. Scale bar = 5 mm. Lower Devonian of Clocksbriggs Quarry, Forfar, Scotland (BMNH Specimen V.58047). c. *Gosslingia breconensis* Heard, showing the sporangia attached laterally along the length of the stem. Scale bar = 10 mm. Lower Devonian (Pragian) of Brecon Beacons Quarry, Wales (NMW Specimen 69.64.G2). From Cleal & Thomas (1999).

Not all zosterophylls had naked stems. Many Early Devonian fossil floras yield stems covered with spines or short leaves. Some belonged to a plant known as *Psilophyton* which is now included in the group known as the trimerophytes (see below). However, others bore bivalved sporangia very similar to *Zosterophyllum* and clearly belong to the zosterophylls. These spiny/leafy zosterophylls are now known as *Sawdonia* (Fig. 4.10b).

The zosterophylls were important components of Early Devonian vegetation, but rapidly declined in numbers and became extinct in Late Devonian times. They were nevertheless most probably ancestors to one of the major groups of Palaeozoic plants, the lycophytes, a group that dominated many Carboniferous and Permian habitats (see Chapter 5).

Trimerophytes

This was the second major group of plants that probably evolved from the rhyniophytes in Early Devonian times (Fig. 4.6). These plants retained a number of features of their rhyniophyte ancestors, such as stems that were naked or only covered by small spines, and had terminally borne sporangia (Fig. 4.12). However, the trimerophytes were morphologically more complex and larger plants than most rhyniophytes, and clearly represented a major evolutionary advance.

Trimerophytes could be at least 0.6 m high (e.g. *Psilophyton forbesii* from Lower Devonian rocks of North America). To provide support for these much larger stems, they needed a stele that was more robust than in the rhyniophytes, accounting for about a quarter of the volume of the stem. The stems branched dichotomously or laterally (monopodially) according to the species, similar to the stems of the rhyniophytes. In certain parts of the plants, however, there were clusters of axes that forked in a more regular dichotomous pattern and which had sporangia at their ends. The trimerophyte sporangia, unlike those of the rhyniophytes, had a clearly developed structure (dehiscence slit) along their length to release the spores on maturity.

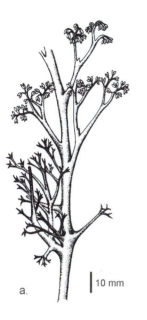

Fig. 4.12 Trimerophytes.
a. Reconstruction of part of a shoot of the Early Devonian trimerophyte *Psilophyton*, showing the characteristic clusters of sporangia. Drawing by A. Townsend, based on the work of H. P. Banks and his colleagues. b. A fertile shoot of *Psilophyton crenulatum* Doran from the Lower Devonian (Emsian) of New Brunswick, Canada. This compressed shoot was extracted by dissolving away the rock matrix with hydrofluoric acid, to show the three-dimensional configuration of the sporangial clusters, and of the small spines attached to the stem. University of Alberta Palaeobotanical Collections, Edmonton, Canada (Specimen S7109). Photo by J. B. Doran, supplied by P. G. Gensel.

a.   10 mm

5 mm   b.

More significantly, the sporangial clusters are the oldest-known examples of complex plant organs with a fixed (determinate) configuration, a growth pattern that we believe led to organs such as leaves and synangia in more advanced plants. Presumably, the grouping of the sporangia into clusters on particular parts of the plant helped maximise their ability to release and disperse the spores.

As with the zosterophylls, the trimerophytes became extinct in Late Devonian times. However, we believe that they were probably ancestral to three of the major groups of vascular plants: the ferns, sphenophytes and progymnosperms. As the progymnosperms were probably the ancestor to all later seed-plants (see next section), we have within the trimerophytes the origin of most of the higher plants living today.

Progymnosperms

By Middle Devonian times, land plants were becoming quite substantial in size, some were reaching 1 m or more tall. However, their basic stem-anatomy still limited their overall growth. The first plants to successfully overcome the anatomical limitations on overall size were the progymnosperms, through developing the process of secondary growth. This is where a lateral meristem (also known as a vascular cambium) is located along the length of the stem between the primary xylem and primary phloem. This continues to divide, producing extra xylem to the inside and extra phloem to the outside of the stem. The resulting secondary xylem (and to a lesser extent secondary phloem) allows the plant to increase significantly the girth of its stems with mechanically strong tissue (wood). This increasing diameter can split the outermost tissues of parenchyma and older phloem on the outside of the stem. To prevent exposure and protect these tissues, the stems form a protective layer that we call bark. The increase in girth is accompanied by an increase in height, eventually giving rise to trees.

Devonian fossilised wood has been known since the nineteenth century (e.g. Fig. 4.13c), but for many years it was not known which plants had produced it. Among living plants, wood only occurs in conifers and angiosperms,

a.

b.

c.

Fig. 4.13 Devonian progymnosperms. a. *Protopteridium thomsonii* (Dawson) Kräusel and Weyland. Scale bar = 5 mm. Sandwick Fish Bed (Middle Devonian), Bay of Skaill, Orkney, Scotland (BMNH Specimen V.9425). b. *Archaeopteris roemeriana* (Göppert) Lesquereux. Scale bar = 5 mm. Upper Devonian (Famennian) of Durnal, Belgium (Royal Institute of Natural Science of Belgium, Brussels, Specimen b 2421 a). c. *Callixylon newberryi* (Dawson) Elkins. Scale bar = 500 μm. New Albany Shale (Upper Devonian), southern Indiana, USA (W. S. Lacey Collection, NMW Specimen 87.77G.589). From Cleal & Thomas (1999).

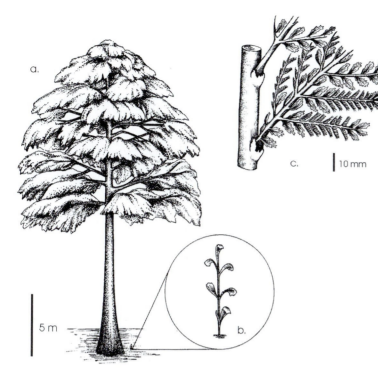

Fig. 4.14 Reconstruction of the *Archaeopteris* tree. a. Complete plant. b. Juvenile plant. c. Shoot showing both foliage and sporangia. Drawing by A. Townsend, based on the work of C. B. Beck and H. N. Andrews.

but no evidence of the leaves of such trees was known from Devonian macrofloras. Eventually, in the 1960s, the American palaeobotanist Charles Beck found examples of woody stems (named *Callixylon*) bearing foliage and spore-producing reproductive structures (already described and named as *Archaeopteris*) indicating that it was a spore-bearing and not a seed-bearing plant (e.g. Fig. 4.13b). From this, Beck developed the concept of the progymnosperms for plants that combined apparently advanced (woody) stem-construction with spore-producing reproductive systems (Fig. 4.14).

Several different types of progymnosperm are now recognised. The oldest, belonging to the order Aneurophytales, are found in Middle and Upper Devonian rocks (Fig. 4.6). For instance, *Protopteridium* is known from northern Scotland (Fig. 4.13a) and was quite a substantial plant probably several metres high. Although a full reconstruction has yet to be made, we know that its branches gave off three-dimensional clusters of axes that appear to have had determinate growth being somewhat reminiscent of the sporangial clusters of

its probable trimerophyte ancestors. Although these clusters of sterile axes retained an essentially three-dimensional branching-pattern, they are regarded as the first stages in the evolution of the photosynthetic organs that we now know as megaphyllous leaves (i.e. relatively large leaves with several veins). The evolution of megaphyllous leaves thus appears to have been a consequence of the development of determinate growth that had originally evolved in the fertile parts of trimerophytes to facilitate spore release.

Only very slightly later, in late Middle Devonian times, do we find more advanced progymnosperms with more modern-looking leaves. These are what we call the Archaeopteridales. The earliest representative of this group is *Svalbardia* (Fig. 4.15), found in Givetian-aged rocks in Spitsbergen and the northern Isles of Scotland (notably Fair Isle). Like the Aneurophytales, the leaves consisted of clusters of axes, but they are now clearly in a flattened (planated) fan-shaped arrangement, which presumably enhanced their efficiency for photosynthesis. During Late Devonian times, an essentially similar plant appeared known as

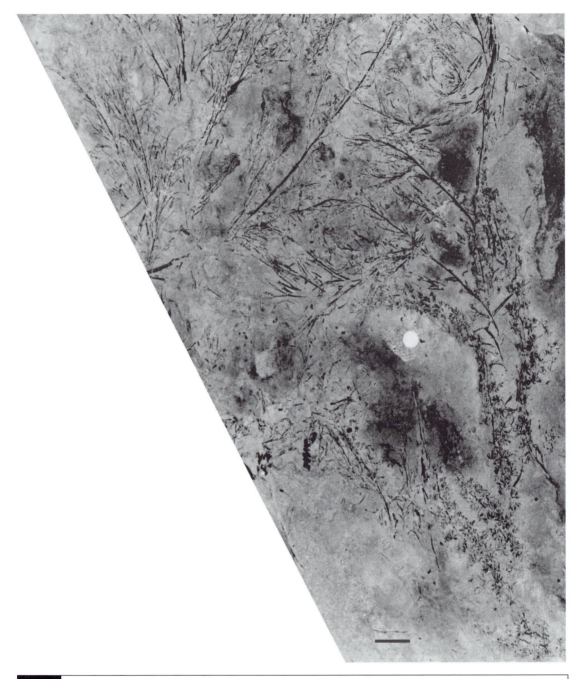

Fig. 4.15 *Svalbardia polymorpha* Høeg. Scale bar = 10 mm. Middle Devonian Series, Planteryggen, Mimerdalen, northwestern Spitsbergen (Paleontologisk Museum, Oslo, Norway, Specimen PA 335). Photo by Franz-Josef Lindemann, Paleontologisk Museum, Oslo, Norway. From Cleal & Thomas (1999).

Archaeopteris (Fig. 4.13b, 4.14), which only differed significantly from *Svalbardia* by its foliage having become at least partially (and in some cases fully) laminated (Fig. 4.15). When these earliest-known modern-looking leaves were first discovered, they were interpreted as being fern fronds, in which the leaflets were arranged in one plane (the generic name implies that it was an ancient fern). However, recent work has shown that the 'leaflets' are helically arranged on the stem and were in fact individual leaves.

The individual sporangia of the Archaeopteridales are similar to those of the Aneurophytales but are borne in a more organised pattern, in two rows along either side of a specialised fertile branch or sporophyll. More importantly, some species of *Archaeopteris* have been shown to be heterosporous, with large megaspores and much smaller microspores, from which female and male gametophytes respectively developed. The evolution of heterospory in the progymnosperms was a critical development in plant evolution, representing what we believe to be the initial step in the evolution of the seed. This is discussed further in Chapter 8.

Recommended reading

Beck (1960, 1988), Chaloner & Macdonald (1980), Edwards (1994, 1996, 1997), Edwards *et al.* (1992), Edwards & Wellman (2001), Gensel & Andrews (1984), Hemsley (1990), Hueber (1968, 1992), Kenrick (1994), Kidston & Lang (1917–21), Niklas & Banks (1990), Remy *et al.* (1993), Thomas & Cleal (2000), Wellman & Gray (2000).

Chapter 5

Lycophytes

Living lycophytes constitute a modest group of small plants that can be easily overlooked in the wild. Many of them (e.g. *Diphasiatrum*, *Huperzia*, *Lycopdium*, *Lycopodiella* and *Selaginella*, see e.g. Fig. 5.1b) have a superficially moss-like appearance, hence their informal name of the club mosses. Others have a more tufted appearance: the quillworts (*Isoetes* and *Stylites*, Fig. 5.1a), and the small *Phylloglossum* has a swollen bulb-like stem and a few relatively long leaves. However, their simple appearance gives a totally misleading impression, as they are what is left of a group of plants that has played an enormously important role in the history of land vegetation. They have the longest fossil record of any living group of plants, extending over 400 million years from at least Early Devonian or possibly late Silurian times through to the present day. For some of this time they were even the dominant plants in the terrestrial vegetation, and in Pennsylvanian times they were the largest known living organisms, forming dense swamp forests over much of the palaeotropical belt.

Extant lycophytes have a simple morphology, with undivided simple leaves (called microphylls). Many can spread vegetatively and all reproduce by spores dispersed from sporangia, which grow in the angle between the microphylls and the stem to which it is attached (this angle is referred to as the axil). The sporophylls may be arranged in fertile zones on otherwise vegetative stems, or in cones on the ends of vegetative shoots. In cones the modified microphylls are referred to as sporophylls. The spores may be all similar (homospores) or of two kinds (megaspores and microspores).

The earliest herbaceous lycophytes

Lycophytes are probably the commonest plant fossils found in Devonian rocks around the world, possibly reflecting a toleration of climatic and other ecological conditions, and/or their better potential for fossilisation. Most are small and described as herbaceous, even though they are known only as fragmentary remains of leafy axes or axes with scars marking the former attachment of leaves. The evidence from petrified small axes is certainly not reliable evidence of them being herbaceous, because these may represent fragments of terminal branches from a much larger plant.

Baragwanathia (Fig. 5.2a) was a vascular plant with leafy shoots that can be closely compared with species of the extant lycophyte genus *Huperzia*. *Baragwanathia longifolia* is the oldest known lycophyte species, being described from two localities in Victoria, Australia. Its exact age has been the subject of much debate for a number of years, but a middle Early Devonian age for both assemblages seems the most realistic. *Baragwanathia* was probably recumbent in growth, branching dichotomously to pseudomonopodially, its leaves were simple, about 30 mm long, and arranged in helices, and it had well-developed adventitious roots arising directly from the leafy stems. All that remains in these adpression fossils is the cuticle, a vascular strand about one tenth the diameter of the stem, and the leaf traces. This suggests that its stem was fleshy and would have easily rotted away. Without the

a.

b.

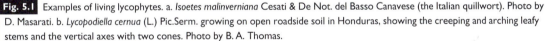

Fig. 5.1 Examples of living lycophytes. a. *Isoetes malinverniana* Cesati & De Not. del Basso Canavese (the Italian quillwort). Photo by D. Masarati. b. *Lycopodiella cernua* (L.) Pic.Serm. growing on open roadside soil in Honduras, showing the creeping and arching leafy stems and the vertical axes with two cones. Photo by B. A. Thomas.

outer layer of thickened, fibre-like cells present in many other lycophytes *Baragwanathia* must have been supported only by turgor pressure. Fertile specimens are uncommon, but the sporangia, when present, are borne in the axils of unmodified leaves in fertile zones on the stems, and released their small, trilete spores through a slit orientated transversely to the leaf.

Asteroxylon mackei from the Devonian System of the northern hemisphere was very similar to *Baragwanathia*. *Asteroxylon mackiei* (Fig. 5.3) was originally described by Kidston and Lang from the Lower Devonian Rhynie Chert flora in Scotland (see Chapter 4). It was interpreted as being superficially very like the living lycophyte *Huperzia selago* in appearance, with a prostrate, creeping rhizome that produced erect stems clothed in helically arranged microphyllous leaves. *Asteroxylon* had a central vascular strand with xylem that was stellate in cross section, and leaf traces ran from it to the bases of the leaves without actually entering them (Fig. 5.2b). The stem apex was flat with several meristematic cells and in this resembles the apex of living *Lycopodium reflexum* Sw. The plants were homosporous, having large kidney-shaped sporangia that appear to have arisen directly from the stem. The sporangia dehisced apically to release one kind of trilete spore. There has been some debate about the exact botanical affinity of *Asteroxylon*: was it a true lycophyte or a lycophyte-precursor? The real problem is that, although many of these early plants resembled living lycophytes, their sporangia were on the stems and not in the axils of leaves, and therefore more similar to the zosterophylls such as *Zosterophyllum* and

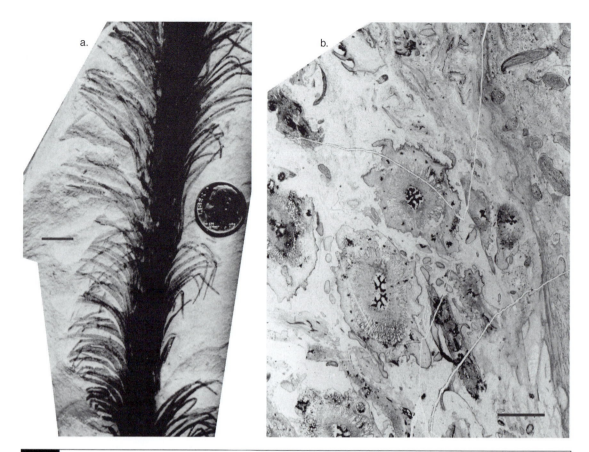

Fig. 5.2 a. *Baragwanathia oblongifolia* Lang & Cookson. Scale bar = 10 mm. From the upper Plant Assemblage (Lower Devonian) of Victoria, Australia (Smithsonian Institution, Washington DC, Specimen USNM 446315). b. Transverse section through a stem of *Asteroxylon mackei* Kidston & Lang showing the star-shaped vascular strand. Scale bar = 5 mm. The small ovals around the stem are sections of its leaves. Rhynie Chert (Early Devonian), near Huntley, Scotland (Hunterian Museum, Glasgow, Specimen 2579). Both from Cleal & Thomas (1999).

Sawdonia. Nevertheless, in our view, the close comparisons of overall morphology and stellar and apical anatomy outweigh any minor differences of reproduction and anatomy, so *Asteroxylon* is surely a lycophyte.

One of the most completely known of the Devonian lycophyte genera is *Estinnophyton* (Fig. 5.4), which had creeping dichotomising stems that occasionally turned upward to bear fertile zones, rather like the living *Huperzia selago* (L.) Schrank & C. Mart. Its stem anatomy was simple, although with a more fragmentary xylem than that of *Asteroxylon*, and its leaves (microphylls) typically divided into five, but sometimes up to seven, apical segments. Each leaf had a little flap of tissue, known as a ligule,

near the base of their upper (adaxial) surfaces and stalked sporangia were borne singly on the upper surfaces of unmodified microphylls. In living lycophytes ligules are restricted to heterosporous genera, but it is often impossible to determine whether or not fossil leafy shoots are ligulate. These plants were very unusual because they appear to be homosporous but ligulate.

It has been suggested that ligules serve some function during the development of the young leaves but this does not explain why some lycophytes have them and some do not. Neither does it explain why the presence of ligules is generally associated with heterospory.

Ligulate lycophytes continued to evolve rapidly during the Carboniferous Period resulting

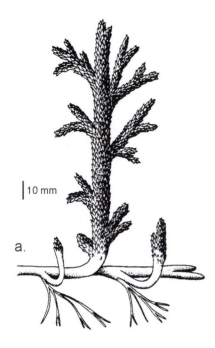

10 mm

a.

5 mm

b.

Fig. 5.3 *Asteroxylon mackei* Kidston & Lang, from the Early Devonian Rhynie Chert.
a. Reconstruction of a whole plant, which was about 0.2 m tall in life, based on the work of D. Eggert.
b. Close-up of the terminal part of a shoot, with attached sporangia. Drawings by A. Townsend, based on the work of W. G. Chaloner.

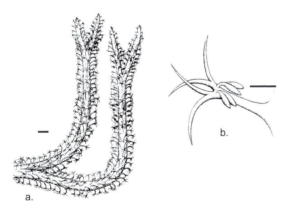

a.

b.

Fig. 5.4 *Estinnophyton*, an Early Devonian lycophyte.
a. Reconstruction. Scale bar = 20 mm. b. Close-up of leaf bearing sporangia. Scale bar = 5 mm. Drawings by A. Townsend, based on the work of M. Fairon-Demaret. From Cleal & Thomas (1999).

Box 5.1 | What is a ligule?

A ligule is a small flat cellular outgrowth from near the base of the adaxial surface of some lycophyte leaves (Fig. 5.5). In the arborescent lycophytes, ligules are sunken in pits, commonly referred to as ligule pits. Ligules do not have a cuticle and are therefore not seen in leaf macerations. In contrast, the cells of the ligule pits have cuticles and are seen in macerations as tubes of square to elongated cells (Fig. 5.16d). This is presumably because they are modified leaf epidermal cells around the base of the ligule that were indented during the development of the young leaf. Ligule pits are open at the base where the ligule was originally attached.

in a number of recognisable forms. The morphologically simplest erect forms, with unknown reproductive organs, had a worldwide distribution in Mississippian times, although they became more restricted during Late Carboniferous and Permian times. Large numbers only survived in the colder northern high palaeolatitudes of Angaraland, where they

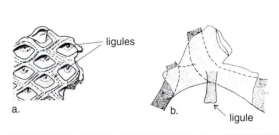

ligules

a.

b.

ligule

Fig. 5.5 Ligules in *Ulodendron majus* Lindley & Hutton. a. Series of leaf cushions showing ligules. b. Cuticle from leaf cushion showing cutinised ligule pit hanging down. Drawings by B. A. Thomas (from Thomas, 1967).

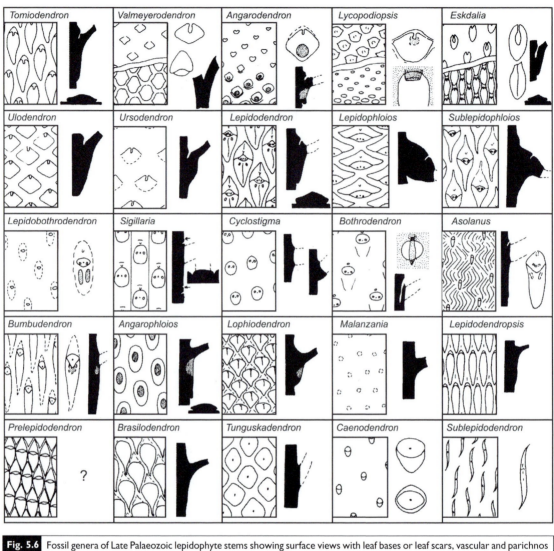

Fig. 5.6 Fossil genera of Late Palaeozoic lepidophyte stems showing surface views with leaf bases or leaf scars, vascular and parichnos marks in the leaf scars and ligule pit apertures, and side views of the 'uncompressed' leaf bases and laminae. Redrawn from Thomas & Meyen (1984).

probably escaped from competition from the more vigorous plants that were evolving in lower palaeolatitudes. These sterile axes are referred to genera generally on morphological characters related to phyllotaxy and leaf morphology, and in some cases features seen after leaf abscission such as swollen leaf bases, leaf scars, ligule pits and infrafoliar bladders (Fig. 5.6). This is a workable system for naming specimens provided its limitations are understood. The systematic and evolutionary relationships of these fragmentary axes are impossible to assess accurately without knowledge of their reproductive organs, so no assumption should be made that putting species in the same genus implies phylogenetic relationships rather than similarity; otherwise it could influence conceptions of plant communities, distribution patterns and palaeogeography.

The beginnings of modern herbaceous lycophytes

A few heterosporous lycophytes, e.g. *Barsostrobus famennensis* Fairon-Demaret, are known from European Late Devonian macrofloras. Heterospory is presumed to have developed in several groups during late Devonian times because the record of fossil spores shows a rapid increase in the number and diversity of presumed megaspores during this period.

In Mississippian times, forms similar to *Selaginella* appeared in Europe. The earliest (*Selaginellites resimus* Rowe) is of Mississippian age from southern Britain, and was herbaceous, isophyllous (that is with only one kind of leaf) and had small terminal cones in which impressions of megaspores can be seen. *Lycopodites* is the name given to small, presumably herbaceous, lycophytes known from the Devonian Period onwards to at least the Early Cretaceous English Wealden. Fertile specimens have globose sporangia on unspecialised sporophylls and spores that are 90–120 μm in diameter. Many of the described species of *Lycopodites* were isophyllous, but *Lycopodites falcatus* Lindley & Hutton had two ranks of larger leaves and two upper ranks of smaller leaves (with twice as many large leaves as smaller ones). Lycophytes with more than one type of leaf are said to be anisophyllous. Anisophylly is not only found in the majority of extant *Selaginella* species, but also in some extant species of *Lycopodium*, e.g. *Lycopodium carolinianum* L.

Further isophyllous forms are known from Pennsylvanian macrofloras, such as *S. fraipontii* (Leclercq), which closely resembles the living *S. selaginoides* (L.) Link in both morphology and internal anatomy. Isolated Triassic cones from Greenland have been included in the same genus, although without any knowledge of their stems the association is rather tenuous.

Anisophyllous lycophytes appeared in the Middle Pennsylvanian Epoch, in the palaeotropical region of Europe and North America, and are so similar to extant anisophyllous species of *Selaginella* that they are included in the same genus. Although the majority of the species are known only from sterile portions of stems, some of them are fertile and heterosporous. Extant anisophyllous species of *Selaginella* have two ranks of larger lower leaves and two ranks of smaller upper leaves, with pairs of each size alternating along the stem. The Carboniferous *Selaginella* species had three ranks, with the third consisting of very small leaves on the underside of the stem, and for this reason they are included as a sixth subgenus of the genus. We have no idea, as yet, when the reduction in the number of ranks from three to two occurred. The first of these fossil *Selaginella* species appeared during the middle of the Middle Pennsylvanian Epoch, in the intramontane Saar-Lorraine basin in central Europe, where the stimulus for their evolution may have been triggered by different competitive forces which favoured a creeping or climbing habit rather than an erect one. By late Middle Pennsylvanian times, such herbaceous forms had spread over an area between Nova Scotia, Canada, to Zwickau in eastern Germany (Fig. 5.7).

Mesozoic anisophyllous species referable to *Selaginella* include *S. anasazia* from the Triassic of New Mexico, USA, *S. hallei* from the late Triassic of Sweden, *S. dichotoma* from the Jurassic of Siberia, *S. dawsonii* from the early Cretaceous of England and *S. nosikvii* from the Cretaceous of the Czech Republic.

At the same time as the lycophytes were diversifying in the equatorial region and expanding their distribution, other genera of lycophytes in northern Angaran Pennsylvanian and Permian floras were showing increasing endemism. There were no truly herbaceous forms in the Angaran floras, only shrubby ones in such genera as *Lophiodendron*, *Tomiodendron*, *Angarophloios*, *Angarodendron* and *Eskdalia*, which formed a sparse 'brush' type of vegetation associated with rivers and lakes (Fig. 5.6). There were also no herbaceous forms in the southern middle or high palaeolatitudes of Gondwanaland, only small trees or shrubs such as the Argentinean *Bubudendron* and *Malanzania*, and the Brazilian *Brasilodendron*.

Fig. 5.7 *Selaginella gutbieri* (Goeppert) Thomas from the Middle Pennsylvanian Series, Zwickau Coalfield, Germany. Scale bars = 5 mm. a. Leafy shoots. b. Terminal cones. Photos by B. A. Thomas.

Fig. 5.8 The Upper Devonian arborescent lycophyte *Cyclostigma kiltorkense* Haughton from Kiltorkan, Kilkenny, Ireland is preserved by chlorite mineral replacement. a. Detached truncated cone segment exposed by the plane of cleavage passing over the ends of the sporangia. Scale bar = 10 mm (Geological Survey, Dublin, Specimen 3075). b. The same specimen photographed under xylene. A megaspore is visible to the right of the cone. Scale bar = 10 mm. c. Leaf scar on the outer surface of the stem. Scale bar = 3 mm. There is a vascular print together with two lateral aerating canals, parichnos, in the upper half of the scar (Geological Survey, Dublin). d. A sporangium full of megaspores. Scale bar = 1 mm (BMNH Specimen V.5934). Photos by W. G. Chaloner. From Cleal & Thomas (1999).

Increase in size and arborescence

Cyclostigma, found in Upper Devonian rocks of the British Isles, China and Japan, was much larger than other Devonian lycophytes, with a main trunk up to 8 m tall and 0.3 m in diameter, and a crown of dichotomising branches (Fig. 5.8). It was an early arborescent lycophyte. Some of the larger branches show a mesh of fine anastomosing lines indicating an expansion of the outer tissues of the stem. The leaves were variable in size, but often attained 150 mm in length on the smaller branches. When they were shed, the leaves left circular or oval scars with a small central vascular scar. *Cyclostigma*'s terminal cones were about 60 mm long and 30 mm in diameter, with a radially elongated sporangium attached to the upper surface of the basal portion of each sporophyll. The sporophyll laminae were about 160 mm long and very similar in appearance to the leaves.

Fig. 5.9 A Pennsylvanian *Lepidodendron* swamp forest of the tropical palaeolatitudes. Drawing by A. Townsend.

rapidly expanded in stem width as they grew upwards until they resembled thick poles covered in leaves. These stems increased in height and girth until they reached their full height and then they divided apically many times. Each successive division produced two smaller branches until the smallest terminal shoots were no more than a few millimetres thick. Other arborescent lycophytes grew lateral branches on their main trunks resulting in more irregular looking plants. Some of these lateral branches were shed leaving large round or oval scars that are usually called 'ulodendroid scars'. In all cases, although the branches underwent some secondary thickening, the narrow stems never expanded to the size of the larger stems. Therefore, the diameter of a lycophyte stem cannot be used as a measure of its age, as can be done with living conifer and angiosperm trees, it only refers to its relative position on the mature plant.

The unusual and rapid growth pattern of these arborescent lycophytes was only possible because they had a support system of thickened cells in the outer parts of the stem, rather than through increasing the size of their central wood as in conifer and angiosperm trees (Fig. 5.11a). The growth pattern, especially the expansion in girth of the trunk and the largest branches, produced a series of abscissions of leaves, 'bark', shoots and fructifications. Consequently, these parts became fossilised as isolated organs, and thus are treated as separate fossil genera for taxonomic purposes.

The arborescent lycophytes reached their peak in the Late Carboniferous (Pennsylvanian) coal-forming swamps and *Lepidodendron* is the generic name that is given to the commonest of the adpression stems. Anatomically preserved stems can be assigned to a number of different genera that cannot all be reliably related to adpression species. *Lepidodendron*'s stems are covered in leaves that have longitudinally elongated swollen bases, called leaf cushions, which remain after the main part of the leaves (the laminae) have been shed (Figs. 5.12, 5.17a). These leaf cushions are known to have had many stomata suggesting that they enabled the stems

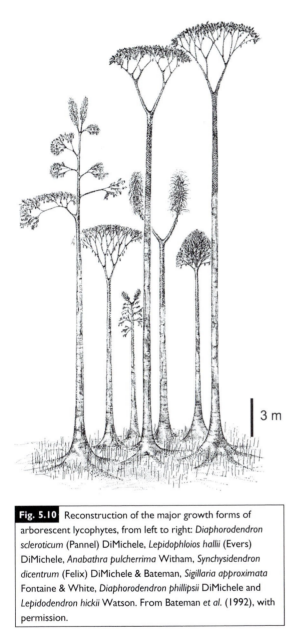

3 m

Fig. 5.10 Reconstruction of the major growth forms of arborescent lycophytes, from left to right: *Diaphorodendron scleroticum* (Pannel) DiMichele, *Lepidophloios hallii* (Evers) DiMichele, *Anabathra pulcherrima* Witham, *Synchysidendron dicentrum* (Felix) DiMichele & Bateman, *Sigillaria approximata* Fontaine & White, *Diaphorodendron phillipsii* DiMichele and *Lepidodendron hickii* Watson. From Bateman *et al.* (1992), with permission.

During Late Palaeozoic times, the average size of the lycophytes steadily increased with some, such as *Lepidodendron*, growing to about 50 m in height with crowns of leafy branches (Figs. 1.8, 5.9). However, even though the arborescent lycophytes resembled trees in appearance, they had rather different growth patterns (Fig. 5.10). In many of them, the young plants

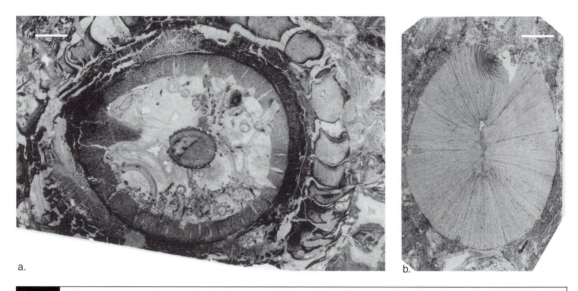

Fig. 5.11 a. Transverse section through a *Lepidophloios* stem. Scale bar = 5 mm. Upper Foot Mine Seam (Lower Pennsylvanian Series), Shore, Lancashire, England (NMW Specimen 21.77.15). b. Transverse section through the woody cylinder of a *Stigmaria ficoides* Brongniart. Scale bar = 5 mm. Halifax Hard Coal (Lower Pennsylvanian Series), Yorkshire, England (NMW Specimen 60.538.9). From Cleal & Thomas (1999).

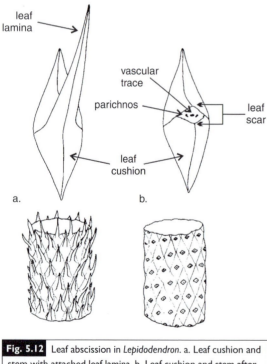

Fig. 5.12 Leaf abscission in *Lepidodendron*. a. Leaf cushion and stem with attached leaf lamina. b. Leaf cushion and stem after lamina abscission. Redrawn from Thomas & Spicer (1987).

to remain photosynthetic. Leaf scars, marking the abscission points of the leaf laminae are left in the centre of the cushions, each with three smaller prints. The central print marks the position of the vascular bundle and the two lateral ones (known as parichnos) the positions of aerating canals that originally ran through the leaves into the outer tissues of the stem most probably facilitating gaseous exchange. Lower down on the cushion surface there are sometimes two larger parichnos, with one on either side of the central line. Above the leaf scar and sometimes adjacent to its upper angle is another small print marking the entrance to a small pit in the bottom of which was attached a ligule. The terminal shoots of *Lepidodendron* trees did not shed their leaves and are notoriously difficult to distinguish. As a result they are usually identified as one of only four species. This is clearly a taxonomic anomaly considering there is a much larger number of species of stems with recognisable leaf cushions. Future studies involving epidermal studies may eventually resolve this problem.

Box 5.2 | Identifying *Lepidodendron* barks, stems and leafy shoots

Species distinction in *Lepidodendron* is through the overall shape of the cushions and their leaf scars, surface markings on the cushions, epidermal details, the position of the ligule pit aperture and the presence or absence of external parichnos (Fig. 5.13). Size is not diagnostic because the main axis and the largest stems have larger cushions than the smaller branches. The cushions may be almost isodiametric to several times longer than broad. Leaf scars are diamond-shaped and may be isodiametric or broader than long. Their positions relative to the cushion length can help in species recognition. The cushion may, or may not, have a central keel above and/or below the leaf scar. The lateral lines that usually run from the edge of the leaf scar to meet the cushion edge demarcate the upper and lower cushion surfaces (effectively the adaxial (upper) and abaxial (lower) surfaces of the leaf lamina), but how far down the length of the cushion they run to meet the edge of the cushion varies between species. The two areas of cushion surface may be similar in appearance, or the upper surface may be finely striated. A tiny scar, marking the opening of the ligule pit, may be adjacent to the upper angle of the leaf scar, or some distance above it. External infrafoliar parichnos may, or may not, be present. Epidermal studies can be a great help in species identification.

It is usually difficult to assign leafy shoots to a morphospecies that has been established on the bark of the main trunk of a *Lepidodendron* tree. Consequently, a separate set of morphospecies has been developed for such shoots, based on the angle of attachment of the leaf to the stem, the curvature of the leaves and the breadth of the base of the leaves (Fig. 5.14).

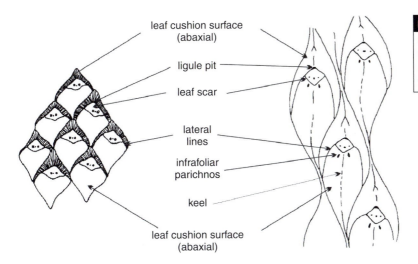

leaf cushion surface (abaxial)
ligule pit
leaf scar
lateral lines
infrafoliar parichnos
keel
leaf cushion surface (abaxial)

Fig. 5.13 The characters of *Lepidodendron* leaf cushions useful in identifying species. Drawing by D. M. Spillards. From Cleal & Thomas (1994).

While *Lepidodendron* is one of the best known of the Carboniferous arborescent lycophytes, there were other genera represented in the Coal Measures forests. *Lepidophloios* and *Sublepidophloios* with their distinctive downward bulging leaf cushions seem to have favoured the wetter parts of the forests (Fig. 5.15). *Sigillaria* is the name given to a type of stem in which the leaf cushions

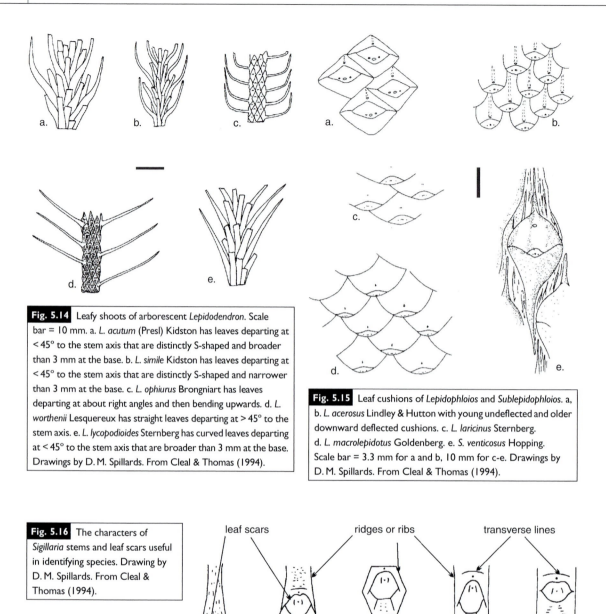

Fig. 5.14 Leafy shoots of arborescent *Lepidodendron*. Scale bar = 10 mm. a. *L. acutum* (Presl) Kidston has leaves departing at <45° to the stem axis that are distinctly S-shaped and broader than 3 mm at the base. b. *L. simile* Kidston has leaves departing at <45° to the stem axis that are distinctly S-shaped and narrower than 3 mm at the base. c. *L. ophiurus* Brongniart has leaves departing at about right angles and then bending upwards. d. *L. worthenii* Lesquereux has straight leaves departing at >45° to the stem axis. e. *L. lycopodioides* Sternberg has curved leaves departing at <45° to the stem axis that are broader than 3 mm at the base. Drawings by D. M. Spillards. From Cleal & Thomas (1994).

Fig. 5.15 Leaf cushions of *Lepidophloios* and *Sublepidophloios*. a, b. *L. acerosus* Lindley & Hutton with young undeflected and older downward deflected cushions. c. *L. laricinus* Sternberg. d. *L. macrolepidotus* Goldenberg. e. *S. venticosus* Hopping. Scale bar = 3.3 mm for a and b, 10 mm for c-e. Drawings by D. M. Spillards. From Cleal & Thomas (1994).

Fig. 5.16 The characters of *Sigillaria* stems and leaf scars useful in identifying species. Drawing by D. M. Spillards. From Cleal & Thomas (1994).

leaf scars   ridges or ribs   transverse lines

ligule pit   lateral lines   basal lines

appear arranged in vertical rows, even though the leaves were formed in spirals from apical meristematic growth (Figs. 5.15, 5.17b). *Sigillaria* was probably a smaller plant than *Lepidodendron*, with an apex that only branched a few times when the trunk reached its maximum height. These plants probably favoured the marginal, slightly drier parts of the swamps.

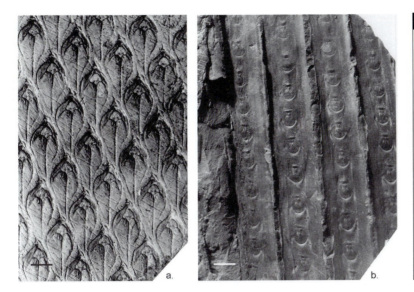

Fig. 5.17 Typical surface features of arborescent lycophytes from the Pennsylvanian palaeotropical wetlands. Scale bars = 10 mm. a. *Lepidodendron aculeatum* Sternberg from the Middle Pennsylvanian Radnice Member, Bohemia, Czech Republic (Prague Museum, Specimen CGH 659 – holotype of *Lepidodendron obovatum* Sternberg, a species now regarded as a later synonym of *L. aculeatum*). b. *Sigillaria nortonensis* Crookall from the Middle Pennsylvanian Coal Measures of the Bristol and Somerset Coalfield (NMW Specimen 90.9G3). Photos by B. A. Thomas.

Box 5.3 | Identifying *Sigillaria* barks and stems

Most species of *Sigillaria* have their leaf scars secondarily arranged into vertical lines in raised ridges or ribs separated by grooves (Fig. 5.15). The ribs may be straight, wavy or broken into hexagons. Others retain their scars on slightly raised areas, similar in some ways to *Lepidodendron* leaf cushions. Species are distinguished by the shape and size of their scars (relative to the width of the ridge), whether or not there are transverse lines above the scars and basal lines below them extending downwards from the lower sides of the leaf scars, and if so how far down they extend towards the next leaf scar. Other useful characters are the relative position of the ligule pit apertures to the upper edge of the leaf scar and markings on the ridges above and below the scars.

Cuticles and paper coal

Cuticular studies on fossil lycophytes have revealed many features. Epidermal details of leaf cushions now help distinguish species of stems, stomatal guard cells are now known to be sunken in pits (Fig. 5.18a,b) and ligule pits have been shown to be cuticle-covered (Fig. 5.18d). Occasionally the preservation process has been slow which has resulted in extensive decay of the plant tissue just leaving layers of plant fossil material that are little more than cuticle. These layers are referred to as Paper Coals and the two best known examples are the Jurassic Paper Coal of Roseberry Topping in Yorkshire, England, consisting mostly of *Pachypteris* gymnosperm leaves, and the Russian

Mississippian Paper Coal of the Moscow Basin consisting entirely of lycophytes. These Moscow Paper Coal lycophytes (Fig. 5.18c) were first described in 1860 as *Lepidodendron*, but have since been referred to the lycophyte genus *Eskdalia*. These slabs of cuticle are from stems of these lycophytes and the oval holes mark the attachment points of leaves. Epidermal cells of the stem cuticle are clearly shown but there are no stomatal guard cells. The small sock-like structures hanging down from the upper angles of the holes are the cutinised inner walls of ligule pits. Fragments of thinner cuticle found attached to the edges of the holes are portions of cuticle from the basal portions of the leaves, and on them can be seen epidermal cells and pairs of stomatal guard cells.

Fig. 5.18 a. *Lepidodendron dichotomum* Sternberg showing epidermal cell outlines and oval stomata from the cushion surface below the leaf scar. The thickened areas alongside the guard cells are cuticles lining the stomatal pits. Scale bar = 20 μm. b. Vertical section through the epidermis and subepidermal layers of *Lepidophloios fuliginosum* from Shore, Lancashire, England, showing guard cells at the bottom of pits. Scale bar = 50 μm. c. Lycophyte papercoal cuticle of *Eskdalia* from the Mississippian Subsystem of the Moscow basin. Scale bar = 250 μm. d. Cuticle from the top of the leaf base of *Ulodendron majus* Lindley & Hutton from the Ayrshire Coalfield showing cuticles from the bases of the leaves and the ligule pit. Scale bar = 500 μm. For further information on lycophyte cuticles see Thomas (1966–1977).

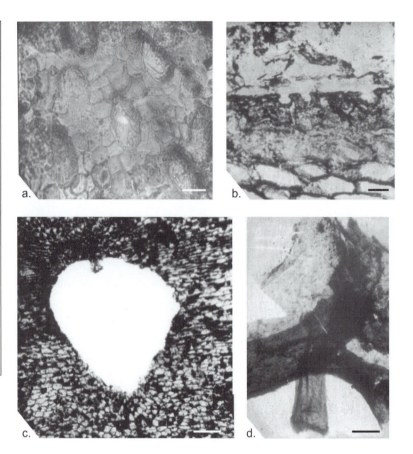

Rooting structures

The arborescent lycophytes with their tall axes and large aerial branching crowns clearly needed an extensive rooting system that would provide for both the physiological needs of the plants and stability in the soft sediments in which they grew. These rooting bases are given the generic name *Stigmaria* and, because there are virtually no characters that can be used to distinguish the stigmarian bases of different species of parent plants, nearly all of them have the same specific name *Stigmaria ficoides*. These rooting bases spread out more or less horizontally from the bases of the main stems and branched dichotomously. As they grew, the older parts increased in girth, partially through secondary xylem production, but mainly by the formation of additional cortical tissue (Fig. 5.11b). Each growing stigmarian apex was a rimmed depression with a protective plug of parenchymatous tissue.

Although these rooting systems have the appearance of roots, their anatomy and morphology is intermediate between stems and true roots, and they are called rhizophores. True roots grew from *Stigmaria* and radiated out from the rhizophore in all directions, reaching up to 0.5 m or more in length when fully grown (Fig. 5.19). The more vertical ones possibly projected out of the sediment into the overlying water. Each root had a large central canal, which provided an internal pathway for gases and possibly stimulated gaseous exchange with the surrounding waterlogged sediments and the overlying water. As the stigmarian axes expanded through secondary growth, the older roots were shed leaving characteristic circular scars on their surfaces.

Many of these stigmarian bases have been preserved as casts in sandstone. After the plants died and their aerial parts disintegrated it was possible for the rotting stigmarian axes to be infilled with fine sediment. In time, this would

a.

b.

Fig. 5.19 a. Stigmaria ficoides Brongniart, from above the Pentre Coal, Gilfach Goch, Wales, showing a small section of a narrow axis to the left and the true roots attached at right angles to the right. Scale bar = 10 mm. The attachment points of the roots can be seen as circular prints all over the stigmarian axis (NMW Specimen 22.113G.97). b. Model of the apical part of a *Stigmaria* showing the roots emerging at right-angles. Modelled by A. Townsend. Both from Cleal & Thomas (1999).

a.

b.

Fig. 5.20 Basal parts of Pennsylvanian arborescent lycophytes. a. Stigmarias in growth position in Victoria Park, Glasgow. Photo by B. A. Thomas. b. A complete stigmarian base in the Manchester Museum, England. Photo by J. Watson.

harden, producing a cast. Manchester Museum has the specimen described by the Victorian palaeobotanist W.C. Williamson in his original work on *Stigmaria* (Fig. 5.20b). It was discovered by quarrymen working near Bradford in 1886 and moved to Manchester at Williamson's own expense. The diameter of the stem and the spread of its rhizophores are 1.3 m and 9.0 m, respectively. The largest axes are 0.5 m in diameter and after two dichotomies the most distant preserved axes are 50 mm in diameter. The best known preserved stigmarian apex is 50 mm in diameter, from which it can be deduced that the Manchester *Stigmaria* is virtually complete.

Many other stigmarias have been found since then in quarry excavations and in eroding sea cliffs, but they are rarely as well-preserved as Williamson's specimen. The two best known sites are the small grove preserved at Victoria Park in Glasgow, Scotland (Fig. 5.20a), and the many horizons of fossil forests at the famous coastal site at Joggins in Canada (Fig. 3.11a). From sites such as these, it has been possible to estimate that there were approximately 1500–2000 of these large tree-like lycophytes per hectare in the Pennsylvanian forests, which is significantly more than the density of trees found in today's tropical rain forests.

Reproduction

These arborescent lycophytes all reproduced by producing cones with a basic morphology of helically arranged sporophylls, each having a single sporangium attached to their upper surfaces. Most cones are found detached, so once again we give them their own system of nomenclature, even if we have knowledge of some of their parent plants. Fossil genera are distinguished by a combination of sporophyll shape and sporangial spore content. Their cones were formed terminally on leafy shoots. Some were bisporangiate with both megasporangia and microsporangia, while others had only megasporangia or microsporangia.

Lepidodendron bore bisporangiate cones (called *Flemingites*) that varied in size from about 0.1 to 1.0 m in length, on the ends of its narrowest shoots (Fig. 5.21). Such cones have been found permineralised in coal balls, giving us details of their anatomy, and as adpressions, allowing their spores to be recovered by acid maceration. Both types of cone had helically arranged sporophylls that were attached at approximately right angles to the cone axis. This 'horizontal' part of the sporophyll is called the pedicel and had an elongate sporangium attached to its upper (adaxial) surface. A ligule in a pit was situated distally to the sporangium. Beyond the sporangium and the ligule, the sporophyll turned abruptly upwards as the leaf-like lamina. Sometimes there was a downward projecting small 'heel' below the leaf lamina. Individual sporangia produced only one kind of spore and usually it was the more apical ones that were microsporangiate. The megaspores were usually quite distinct and they can be used to help distinguish species of otherwise similar cones. Most were about 1 mm in size, spherical or with a resemblance to an inflated sack with a triangular projection where the three contact faces were located. The body of the spores may have been smooth or covered in spines, being closely comparable to the dispersed spore genera *Lagenoisporites* or *Lagenicula*. The microspores were about 25–35 μm in diameter, trilete, with smooth contact faces and granular distal surfaces. Because there were thousands of microspores in each sporangium, there was some natural variation amongst them. Nevertheless, they all fit within the limits of the dispersed spore species *Lycospora* and with careful observation different populations can be identified in different species of cone.

Cones produced by the arborescent lycophytes called *Sigillaria* are called *Sigillariostrobus*, which can be distinguished from those included in *Lepidostrobus* and *Flemingites* by their more triangular sporophyll shapes and the possession of a peduncle (Fig. 5.23). *Sigillariostrobus* cones are all monosporangiate and produced only megaspores or microspores.

Other arborescent lycophytes developed a reproductive strategy that was rather different; the best known of these being the plants that had *Lepidophloios* branches and *Lepidocarpon* cones. *Lepidophloios*, like *Sigillaria*, produced two kinds of cones, but each sporangium in the megasporangiate cones contained only a single functional megaspore and the sporangium was partly enclosed within lateral extensions of the sporophyll. This sporophyll was, therefore, partly

fulfilling the protective role of the integument in a gymnosperm seed (discussed in Chapter 8). Dispersal was achieved by the release of the entire sporophyll unit (called *Lepidostrobophyllum* when found isolated) from the cone rather than by shedding the individual spores (Fig. 5.22). The microspore-producing cones, called *Lepidostrobus*, are only distinguishable from *Flemingites* by their spore contents. Although there is no suggestion that these arborescent lycophytes were related to gymnosperms, it is interesting that they had

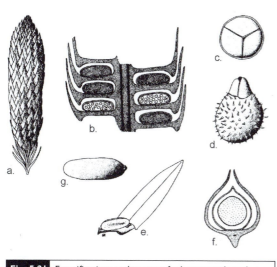

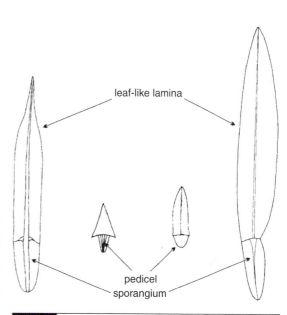

Fig. 5.21 Fructifications and spores of arborescent lycophytes. a-d.; a. Reconstruction of *Flemingites*. b. Longitudinal section through a *Flemingites* cone showing sporophylls, and adaxially attached megasporangia and microsporangia. c. Microspore (about 25 μm in diameter) comparable to the dispersed spore genus *Lycospora*. d. Megaspore (about 1 mm in diameter) comparable to the dispersed spore genus *Lagenicula*. e. Reconstruction of a *Lepidostrobophyllum* dispersed sporophyll. f. Cross section through *Lepidocarpon*. g. The single tetrad of megaspores from a *Lepidocarpon* sporangium. The larger fertile spore is comparable to the dispersed spore genus *Cystosporites*. The three aborted megaspores are on the right. All taken from Thomas & Spicer (1987).

Fig. 5.22 Characters of *Lepidostrobophyllum* sporophylls useful in identifying species. The middle figures are viewed from the lower surface, the outer two from above.

Box 5.4 | Identifying *Lepidostrobophyllum* sporophylls

These sporophylls were shed naturally from the parent cones and are widespread in Carboniferous tropical macrofloras. Most sporophylls are found flattened, although in life the pedicel and lamina are almost at right angles to each other. They have a basal (proximal) pedicel by which they are connected to the cone axis, usually with an attached sporangium, and an apical (distal) leaf-like lamina. The sizes of sporophylls vary within a single cone with the most terminal sporophylls usually sterile and much smaller. The overall shape and apical shape of the lamina and the distinctness of the central vascular strand are the best characters used for species distinction. The relative size of the sporangia can help in species determination (Fig. 5.22).

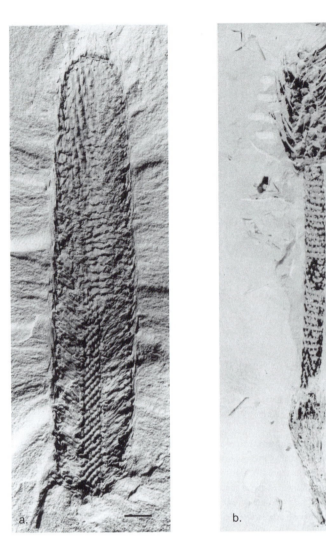

Fig. 5.23 Arborescent lycophyte cones. Scale bars = 10 mm. A. *Lepidostrobus* from the South Wales Coal Measures at Abercarn, Gwent, Wales (NMW Specimen G.1007). b. *Sigillariostrobus* from Seam F (Middle Pennsylvanian Series) of Dortmund, Germany (Museum fur Naturkunde, Berlin, Specimen 2690). These cones shed their sporophylls from the base upwards exposing the cone axis with its spiral of sporophyll scars. The peduncle is the thicker basal portion. From Cleal & Thomas (1999).

started to develop a broadly similar reproductive strategy.

After the giants

Tectonic activity towards the end of the Carboniferous Period caused the immense palaeotropical coal swamps to start drying out and the arborescent lycophytes, together with several other plant groups, went into decline and ultimately extinction. *Lepidodendron* and *Sigillaria* lasted a little longer in the more inland and isolated basins, but had died out here as well by the end of the Carboniferous Period.

Further east, especially in China, arborescent lycophyte forests continued to flourish and even expand during early and middle Permian times, but towards the end of the Permian Period they had started to disappear even here. The problem for these large plants seems to have been that they were too well adapted to a specific habitat – wetlands in which the substrate was both very soft and very acidic. When these habitats went into decline, so did these plants, as they could grow nowhere else.

Plants like the smaller, sub-arborescent *Omphalophloios* that were 2–6 m tall, lasted until the end of the Carboniferous Period (Fig. 5.24). The even smaller *Pleuromeia* is known from

Fig. 5.24 Reconstruction of the smaller arborescent lycopsid *Omphalophloios* (up to about 6 m tall) in the Palaeobotanical Museum in Cordoba, Spain. Photo supplied by R. H. Wagner (see Wagner *et al.*, 2003 for further details).

many Early to Middle Triassic localities in the mid-palaeolatitudes of the northern hemisphere (Fig. 5.25). Its vertical unbranched stems ranged in height from 0.2 to 3.0 m and were covered in simple lanceolate leaves that could be up to 100 mm long on the largest stems. These leaves were shed from the basal regions as the stem grew. Eventually a terminal cone was formed with rounded sporangia attached to the upper surface of distinctive broad ovate sporophylls. The species were all heterosporous although individual cones were monosporangiate, containing either microspores or megaspores. The smallest species had simple bulbous rooting bases while the largest had four-lobed bases reminiscent of *Stigmaria*.

Pleuromeia was one of the few plant genera to flourish after the devastating Permian/Triassic extinction event that caused so many of the dominant Palaeozoic plant groups to disappear (see Chapter 11). There are several localities where *Pleuromeia* has been found in stands suggesting that it grew as the dominant vegetation over large areas within coastal plain and deltaic drainage basins. There is still some debate over whether some such stands may have grown in brackish environments such as marine embayments.

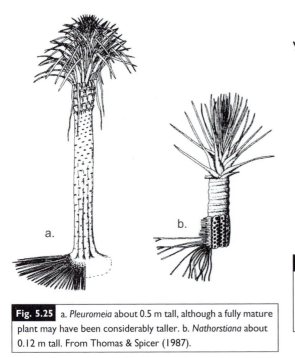

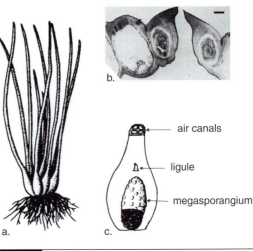

Fig. 5.25 a. *Pleuromeia* about 0.5 m tall, although a fully mature plant may have been considerably taller. b. *Nathorstiana* about 0.12 m tall. From Thomas & Spicer (1987).

Fig. 5.26 a. *Isoetes echinospora* Durieu plant, about 80 mm tall. b. Section through stomata of *Isoetes histrix* Bory. Scale bar = 10 μm. c. The inner side of the lower part of a megasporophyll with its megasporangium. From Thomas & Spicer (1987) (a and c) and Thomas & Masarati (1982) (b).

Plants with short corm-like bases and long leaves similar to the growth habit of *Isoetes* are known from Upper Triassic rocks, and even if most of their remains are fragmentary, they are much more common in Late Cretaceous and Cenozoic floras. These are all referable to the genus *Isoetites*.

An understanding of *Omphalophloios* and *Pleuromeia* is central to our interpretation of the evolution of modern *Isoetes*. Genera such as the Triassic *Takhtajanodoxa* and *Isoetes*-like *Isoetites*, and the Cretaceous *Nathorstiana*, could be considered as parts of a reduction series in the size of the lycophytes. Indeed, in 1956, Mägdefrau proposed a simple reduction series from the Carboniferous lepidodendraleans through the Triassic *Pleuromeia* and the Mesozoic *Nathorstiana*, to living *Isoetes* (Fig. 5.26). Documented changes among the Isoetaleans which probably gave rise to the modern *Isoetes* include the reduction of the axis, the development of sunken sporangia with a covering velum, a change from trilete to monolete microspores, and the development of a swollen ligule base (glossopodium). Other theories suggest that the initial member of the reduction series is more likely to have been the subarborescent lycophyte *Omphalophloios* which hardly differed from *Pleuromeia*. The problem for such a reduction series is that both *Pleuromeia* and *Isoetites* were living at the same time in Mesozoic times. So, it is most likely that *Isoetes* is not the result of a single line of reduction, but rather the remnant of a Palaeozoic and Mesozoic mosaic of forms.

Recommended reading

Ash & Pigg (1991), Bateman *et al.* (1992), Brack-Hanes & Thomas (1983), Cleal (in Benton, 1993), DiMichele & Phillips (1985), DiMichele & Skog (1992), Gastasldo (1986), Hopping (1956), Macgregor & Walton (1955), Phillips & DiMichele (1992), Rothwell (1984), Thomas (1966, 1970, 1974, 1977, 1978, 1981, 1997, 2005), Thomas & Cleal (1993), Thomas & Meyen (1984), Wagner *et al.* (2003).

Chapter 6

Sphenophytes

The Sphenophytes (often referred to informally as horsetails) have never been a very diverse group of plants compared to the other spore-bearing plants. Today there is only the one genus, *Equisetum*, with about 20 species (Fig. 6.1). Nevertheless the plants are often very successful growing throughout much of the world in a variety of habitats. They are only absent from Australia, New Zealand (except for some naturalised introductions) and Antarctica.

Horsetail stems are distinctive in appearance, with branches and leaves borne in whorls at the nodes of ribbed erect stems (Fig. 6.2). Their stems are hollow, or are filled in the centre with soft, easily degraded pith. The stems are always solid at the nodes where there is a diaphragm of tissue stretching across the width of the stem. In all living species and some of those found in the fossil record, the leaves are fused basally to form a distinctive sheath around the stems. In contrast, the leaves of most Palaeozoic genera are not fused or are only joined by a very narrow collar of tissue.

Equisetum spreads by an underground rhizome often resulting in swards of the distinctive upright stems (Fig. 6.1) and the evidence suggests that many of the extinct species grew this way. Reproduction in all extant and extinct genera is by spores produced in cones that are borne on the ends of stems. Except in some of the more primitive Palaeozoic forms, the sporangia are attached to the undersides of specialised peltate (mushroom-shaped) sporangiophores. The spores in *Equisetum* and some extinct genera have distinctive strap-like appendages called elaters that

probably help with their dispersal by coiling and uncoiling with changes in humidity (Fig. 6.2). Their prothalli are relatively large and bisexual.

Origin and systematic position of the sphenophytes

There is unequivocal fossil evidence of sphenophytes as far back as Late Devonian times. They probably originated from a group of early fern-like plants known as the Cladoxylales that included plants such as *Calamophyton* (see Chapter 7); a group that itself arose from the trimerophytes. *Calamophyton* and its close allies do have many features in common with the early sphenophytes, such as tufts of sterile axes that resemble leaves, and sporangia borne on the recurved tips of slender axes that are clustered into loose cones.

Sphenophytes have a mode of reproduction that is very similar to that of the lycophytes and ferns, and many botanists still group them all together into the division Pteridophyta. Because this type of reproductive cycle appears to be a feature that has been retained from their Palaeozoic ancestors, it suggests that the fern, lycophyte and sphenophyte lineages were independent at least as far back as the Devonian Period. It is, therefore, more reasonable to assign the three groups to their own separate divisions or phyla. There has been disagreement about what formal name to give the horsetail phylum with different authors referring to it as the Equisetophyta, Arthrophyta or, as we will use here, the Sphenophyta.

Fig. 6.1 A patch of *Equisetum telmateia* Ehrh. spreading by underground rhizomes on a roadside in northern France. The vertical stems are white and the whorled branches are green. Photo by B. A. Thomas.

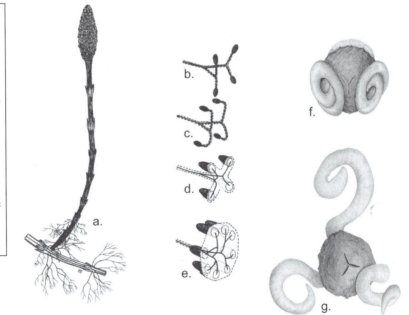

Fig. 6.2 a. *Equisetum*. Portion of the underground rhizome with a fertile aerial stem terminating in a cone. Roots grow at the nodes of the rhizome and the underground portion of the aerial shoot. b–e. Suggested evolutionary stages of the *Equisetum* sporangiophore (e) from terminally fertile primitive appendages (b). Broken lines indicate the appendage outline and solid lines the vascular supply. f–g. *Equisetum* spore with elaters coiled (f) and uncoiled (g). From Thomas & Spicer (1987) (a–e) and Thomas & Cleal (1993) (f, g); the latter are paintings by P. Dean.

Three orders are normally recognised within the sphenophytes: the Pseudoborniales, Sphenophyllales and Equisetales. All the plants included in the first two orders are extinct, while the third includes the family of living sphenophytes, the Equisetaceae, plus those extinct plants included in the two Palaeozoic families, the Archaeocalamitaceae and Calamostachyaceae.

Pseudoborniales

The oldest undoubted example of a horsetail is of Late Devonian age, from Bear Island in the North Atlantic, and Alaska. The plant, called *Pseudobornia*, was probably the size of a small tree, up to 20 m high, with a trunk about 0.6 m

Fig. 6.3 The type specimen of *Pseudobornia ursina* Nathorst showing leaves borne in whorls around the stem. Upper Devonian Series of Bear Island, Arctic Ocean (Swedish Museum of Natural History, Stockholm). From Cleal & Thomas (1999).

wide at the base. The trunk's nodes only gave rise to one or two branches that could be up to 3 m long. These branches bore lateral secondary branches, which themselves bore the ultimate leafy branches. Leaves were borne in distinct whorls of four, forked two or three times near the base, with each segment further divided into pinnae (Fig. 6.3). Cones were borne at the ends of the first order branches, in the upper part of the plant and consisted of alternating whorls of bracts and modified branches (sporangiophores) bearing about 60 sporangia at the ends of recurved axes.

Pseudobornia had clear horsetail characteristics of ribbed stems, whorled leaves and sporangiophores, and sporangia borne in cones. However, there are also features linking it with the cladoxylaleans such as the complexity of the leaves and having its sporangia on recurved axes. It therefore seems to represent an intermediate position between the sphenophytes and their cladoxylalean ancestors, and most palaeobotanists now agree that it is a primitive horsetail and place it in this monotypic order.

Box 6.1 | Identifying *Sphenophyllum* foliage

As with all sphenophytes, *Sphenophyllum* leafy shoots have leaves in whorls around the stem. However, they are easily differentiated from other sphenophytes because the leaves have several veins and are usually wedge-shaped. They were small scrambling plants that mainly spread by vegetative propagation. Consequently, the bulk of the fossils of these plants that we find are of slender, ribbed stems with wedge-shaped leaves borne in whorls. They are identified by the shape, the degree of incision and the characters of the distal margins of the leaves. Fig. 6.4 shows the leaves of some of the more widespread species found in Late Carboniferous macrofloras of Euramerica.

The main difficulty with identifying these fossils is the variation that can occur in leaf shape in different parts of the plant. The leaves shown in Fig. 6.5 were from the lateral, aerial shoots of the plant and are the most reliable ones to use in identification. However, the upright stems that bore these lateral shoots, and the horizontal stems that spread across the ground, had much more deeply incised leaves, that are difficult, if not impossible, to use in distinguishing between species. There are some species (e.g. *S. myriophyllum*) that have deeply incised leaves of their stems which makes identification potentially even more difficult.

The lateral leafy shoots tend to be much more abundant than the upright and horizontal stems. If a reasonable number of specimens are available from a particular macroflora, it is usually possible to distinguish the more abundant lateral leafy shoots from the rarer horizontal and upright shoots with their more deeply incised leaves. In these cases, the species identification is mainly based on the former. However, if only a single specimen is available, and it shows deeply incised leaves, great care has to be exercised, and it is probably wisest to name it merely as *Sphenophyllum* sp.

Sphenophyllales

Plants referable to this order first appeared in Mississippian or possibly very late Devonian times, and then flourished in the Pennsylvanian palaeoequatorial forests. In Permian times, the order extended its distribution and has been recorded from the Amerosinian Palaeokingdom (Fig. 11.4) throughout most of Gondwanaland (mainly India, South America and Africa). Records of Triassic-age sphenophylls are doubtful, and the order almost certainly became extinct at the end of the Permian Period.

The most commonly found genus of this order in the Pennsylvanian palaeotropical floras is the herbaceous *Sphenophyllum* (Figs. 6.4–6.6). These plants had many horsetail-like features, including ribbed stems, whorled leaves, and cones with whorls of sporangiophores and bracts. However, the anatomy of the stems was quite different in having a solid stele that was usually triangular in cross section. The leaves were also more complex than those of any other horsetail, being wedge-shaped with numerous dichotomous veins. The earliest known species had highly dissected leaves, but there was a trend in succeeding species towards entire leaves with more spreading veins. *Sphenophyllum* showed a range of growth forms suggesting that different ecotypes had developed in different habitats. Some are thought to have been components of the swamps, while others that had spines on their stems and hooks (protruding veins) at the tips of their leaves, were probably scrambling plants. These scrambling species may have occupied open, disturbed ground or even formed a low scrub of mutually supportive individuals, thereby filling a similar niche to the living goosegrasses and bedstraws (*Galium*) in today's temperate floras.

The bracts of the cones were an expansion of the sterile basal part of the sporangiophores, rather than independent structures as in the calamite cones. This distinctive feature has

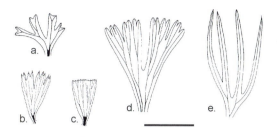

Fig. 6.4 Examples of *Sphenophyllum* leaves. Scale bar = 10 mm. These are identified by the shape, the degree of incisions and the characters of the distal margins. a. *S. trichomatosum* Stur with a length/breadth ratio of about 2, lateral margins more or less concave, distal margins deeply incised with a median cleft > 0.5 of leaf length. b. *S. oblongifolium* (Germar & Kaulfuss) Unger with a length/breadth ratio of 3 or more, median incision < half the length of the leaf, other incisions less, distal margin with drawn-out pointed teeth. c. *S. cuneifolium* (Sternberg) Zeiller with a length/breadth ratio of 3 or more, incisions very small, distal margin with sharply pointed lobes. d. *S. majus* (Bronn) Bronn with a length/breadth ratio of approximately 2, lateral margins straight to convex, distal margin with sharply pointed teeth. e. *S. myriophyllum* Crépin with a length/breadth ratio 3 or more, lateral margins more or less convex, incisions extending for two-thirds or more of leaf length, lobes slender and spreading. Drawings by D. M. Spillards. From Cleal & Thomas (1994).

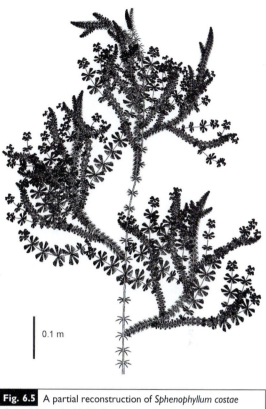

0.1 m

Fig. 6.5 A partial reconstruction of *Sphenophyllum costae* Sterzel from the Middle Pennsylvanian Sydney Mines Formation, Sydney Coalfield, Nova Scotia, Canada. From Bashforth & Zodrow (2007), reproduced with permission.

caused some authors to question whether the Equisetales and Sphenophyllales shared a common ancestor, or whether they evolved independently from the cladoxylaleans. There have even been suggestions that the Sphenophyllales had more in common with the lycophytes than the sphenophytes. Whatever the eventual result of this debate, there is no doubt that the sphenophyllaleans were important components of the Late Palaeozoic palaeotropical vegetation of the Amerosinian Palaeokingdom (Fig. 6.6b).

Archaeocalamitaceae

These sphenophytes became common components of land vegetation during Mississippian times. The stems of the early plants are called *Archaeocalamites* and, like most sphenophytes, are characteristically ribbed with transverse furrows marking the positions of the nodes (Fig. 6.7). Petrified specimens show that there was a well-developed pith cavity along the stem, the first time this horsetail characteristic

is found in the fossil record. Unlike modern sphenophytes, however, many *Archaeocalamites* stems show evidence of secondary wood, indicating that they could have been parts of large plants.

There are a number of primitive characteristics in *Archaeocalamites* which link it with the Pseudoborniales and their cladoxylalean ancestors. The forking leaves somewhat resembled enlarged *Pseudobornia* leaves, but without the ultimate pinnate segments. The ribs were continuous along the length of the stem, differing from the later sphenophytes where the ribs alternated with the furrows at the nodes. The loose cones, usually referred to the fossil genus *Pothocites*, had sporangia in groups of four on the underside of peltate sporangiophores, similar to those found in the later calamostachyacean cones. *Pothocites* usually did not have bracts, making it seem more reminiscent of the modern *Equisetum* cones. It was sometimes

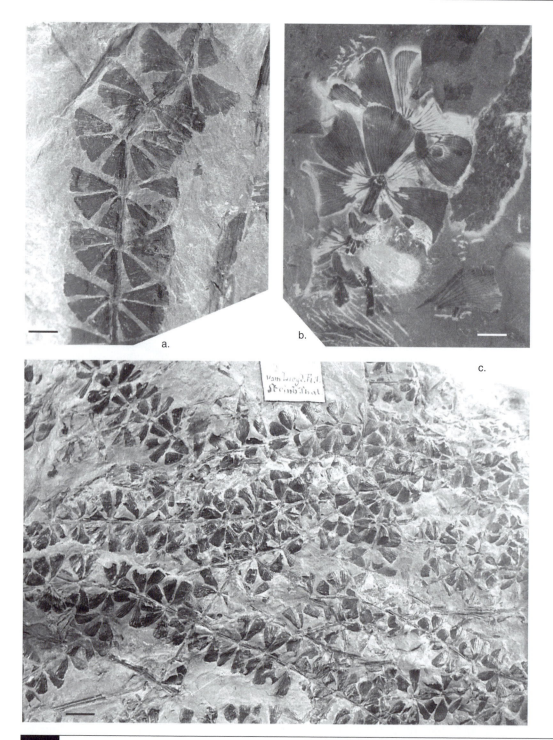

Fig. 6.6 *Sphenophyllum* shoots and foliage. a. *Sphenophyllum emarginatum* Brongniart, Beust Seam (Middle Pennsylvanian Series), Mine Göttelborn, Saarland, Germany (Saarbrücken Mining Museum Specimen C/4060). Scale bar = 5 mm. b. *Sphenophyllum* sp., Lower Permian Series, Simugedong, Taiyuan East Hill, Shanxi, China (NMW Collections). Scale bar = 5 mm. c. *Sphenophyllum emarginatum* Brongniart, mass of shoots, Seam 1 (Middle Pennsylvanian Series), Mine Von der Heydt, Saarland, Germany (Saarbrücken Mining Museum Specimen C/3972). Scale bar = 10 mm. Photos a and c by C. J. Cleal; b from Cleal & Thomas (1999).

heterosporous, but the microspores did not have the distinctive elaters found in both the Calamostachyaceae and modern sphenophytes. This apparent mixture of primitive and advanced features has resulted in *Archaeocalamites* and *Pothocites* being assigned to their own family.

Archaeocalamites was for a long time regarded as a characteristic plant of Mississippian macrofloras, with no known occurrences in rocks of Pennsylvanian age; it was widely believed that they were out-competed by their more

sophisticated descendants the Calamostachyaceae. This view may have to be revised in the light of the recent discovery of Permian fossils that appear identical to *Archaeocalamites* stems. Although the identifications need to be confirmed by the discovery of fertile material, it looks likely that *Archaeocalamites* was only displaced from the wet, lowland habitats, and that it was perfectly well adapted to survive in the drier, extra-basinal areas. This may have consequences for our understanding of the origin of the Equisetaceae, which will be discussed later.

Calamostachyaceae

The sphenophytes reached their maximum physical size and probably diversity in Pennsylvanian times, with the appearance of large plants that we group into the Calamostachyaceae. These plants seem to have mainly favoured the wetter parts of the palaeotropical belt, where they flourished around the fringes of lakes. They appear to have been able to recover from being inundated with mud, growing quickly upwards through the newly deposited sediment (Fig. 6.8). Their fossils are extremely abundant in Pennsylvanian macrofloras of Europe and North America, and are also common in China where they persisted into the Permian Period.

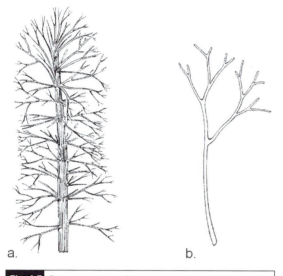

Fig. 6.7 Reconstructions of *Archaeocalamites*. a. A leafy stem. b. A single leaf. From Stur (1875).

Fig. 6.8 Vertical stems of *Calamites* preserved *in situ*. These stems probably grew up through sediment that had inundated the habitat. Bolt Mountain Section, West Virginia, USA. Photo by C. J. Cleal.

Like most other sphenophytes, except *Archaeo-calamites*, most of them spread by a creeping rhizome that produced vertical stems, and secondarily thickening wood enabled them to grow to heights of at least 10 m. They branched at the nodes and these branches would themselves have borne several orders of branches, producing a tree-sized plant that has been described as looking like a 'giant bottle brush' (Fig. 6.9).

Adpressions of the stems of these plants are usually assigned to the fossil genus *Calamites*. These stems had scars indicating the positions of former leaves and sometimes scars where branches were attached. The stems had a central ribbed cavity in the pith, which after death would sometimes become infilled with sediment and the resulting pith casts are also assigned to the fossil genus *Calamites* (Fig. 6.11a). The pattern of the ribs, which became offset at each node, indicated that these plants produced vascular strands to each leaf in a similar way to the living *Equisetum*, and not as in *Archaeocalamites* where there is no such offset. Another characteristic of *Calamites* pith casts is that they narrow markedly near where they are attached to the lower order branches. Stems

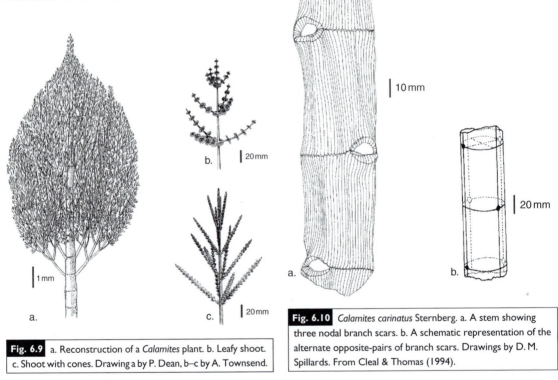

Fig. 6.9 a. Reconstruction of a *Calamites* plant. b. Leafy shoot. c. Shoot with cones. Drawing a by P. Dean, b–c by A. Townsend.

Fig. 6.10 *Calamites carinatus* Sternberg. a. A stem showing three nodal branch scars. b. A schematic representation of the alternate opposite-pairs of branch scars. Drawings by D. M. Spillards. From Cleal & Thomas (1994).

Box 6.2 | Identifying *Calamites* stems and pith casts

Both adpressions and pith casts of the stems are assigned to the morphogenus *Calamites* (Fig. 6.10). Distinguishing morphospecies of *Calamites* is difficult. The characters normally used to differentiate morphospecies are the lengths of the internodes relative to their width, any variability of internode lengths, the distinctiveness of the vertical ribs, the presence, number and positioning of branch scars at nodes. The branch scars may be at every node or a certain number of internodes apart, in opposite and alternate pairs, or in rings that may be complete with distant or touching scars.

Fig. 6.11 *Calamites* stems. a. A pith cast of *Calamites* in growth position. The plant was pushed from the vertical by an inflow of sediment-laden water, which entered and filled the stem cavity. Middle Pennsylvanian Series, Wrexham Coalfield, Wales. b. Transverse section of *Arthropitys* from a thin-section of a coal ball, showing a ring of primary vascular bundles surrounded by radiating tracheids of the secondary wood. The cortical tissues have been lost and the pith cavity invaded by two stigmarian rootlets. Scale bar = 1 mm. Halifax Hard Coal (Lower Pennsylvanian Series), Yorkshire, England (NMW Specimen 60.538.1). Photos by B. A. Thomas.

Box 6.3 | Identifying *calamostachyalean* foliage

The foliage consists of simple linear leaves borne in whorls around the ultimate branches of the plant. One of the commoner leaf-types belongs to the form-genus *Annularia*, in which there can be up to 20 elongate, blade-like leaves in each whorl (Fig. 6.12). The whorls of leaves are often found preserved in about the same plane as the stems suggesting that they were obliquely arranged on the stem. Narrower leaves are called *Asterophyllites* and such leafy shoots often have their leaves extending into the sediment suggesting that they were not obliquely orientated as in *Annularia* (Fig. 6.13 and 6.14b). Characters used to identify morphospecies include the size of the leaf, where along its length it is widest, and the shape of the apex.

---

that are anatomically preserved are referred to the genera *Arthropytis*, *Calamodendron* or *Arthroxylon* on the basis of their xylem organisation (Fig. 6.11b).

Whorls of leaves were retained on the nodes of the ultimate branches. The leaves were simple, with a single longitudinal vein, and were sometimes densely covered with hairs. Stomata

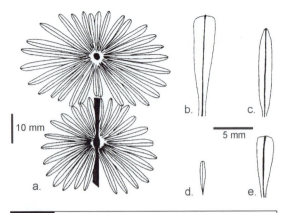

Fig. 6.12 Examples of *Annularia* shoots and leaves. a. *A. stellata* (Sternberg) Wood with oblanceolate leaves up to 75 mm long, widest in the middle. b. *A. mucronata* Schenk with spatulate leaves 12 mm or more long, with mucronate tips, widest above the middle. c. *A. radiata* (Brongniart) Sternberg with leaves 6–15 mm long with acuminate apices, widest in the middle. d. *A. galloides* (Lindley & Hutton) Kidston with leaves 1–4 mm long with blunt apices, widest in the middle. e. *A. sphenophylloides* (Zenker) Gutbier with spatulate leaves less than 12 mm long, widest above the middle. Drawings by D. M. Spillards. From Cleal & Thomas (1994).

were on both surfaces of the leaves. In some cases (*Annularia*) the whorls of leaves were obliquely arranged on the stem, probably oriented towards the prevailing light to optimise their photosynthetic efficiency. *Annularia* leaves were not fused at the base as is seen in the leaf sheaths of modern *Equisetum*, although there may have been a very narrow collar of fused leaf at each whorl, which can be interpreted as an incipient leaf sheath (Fig. 6.14a).

The cones were borne in branching clusters and have been assigned to several genera (Fig. 6.14c–e). All of the cones have alternating whorls of sterile bracts and sporangiophores with each usually having two or four sporangia. All but a few of these cones were homosporous and the spores had elaters as in modern *Equisetum*. The exception is *Calamocarpon* that had sporangia containing just a single megaspore. The suggestion is that *Calamocarpon* might have been developing towards a seed-like reproductive strategy similar to that seen in some Palaeozoic lycophytes (see Chapter 5).

Fig. 6.13 Examples of *Asterophyllites* shoots and leaves. a. *A. equisetiformis* Brongniart with more than 10–25 leaves per whorl; leaves > 10 mm long. b. *A. lycopodioides* Zeiller with less than 6 leaves per whorl; leaves < 3 mm long, attached obliquely to stem. c. *A. grandis* (Sternberg) Geinitz, which has 6–10 leaves per whorl; leaves > 3 mm long. d. *A. longifolius* (Sternberg) Brongniart, which has more than 25 leaves per whorl; leaves > 50 mm long. e. *A. charaeformis* (Sternberg) Unger with less than 6 leaves per whorl; leaves < 3 mm long attached at right angles to stem and strongly arched. Scale bar = 10 mm, except for e where scale bar = 1 mm. Drawings by D. M. Spillards. From Cleal & Thomas (1994).

Fig. 6.14 Foliage and cones of calamite sphenophytes. Scale bars = 10 mm. a. *Annularia stellata* (Sternberg) Wood, Farrington Formation (Middle Pennsylvanian Series), Lower Writhlington Colliery, near Radstock, Somerset, England (NMW Specimen 90.8G.4). b. *Asterophyllites equisetiformis* Brongniart, Farrington Formation (Middle Pennsylvanian Series), Lower Writhlington Colliery, near Radstock, Somerset, England (NMW Specimen 90.8G.3). c. *Palaeostachya* sp., Farrington Formation (Middle Pennsylvanian Series), Lower Writhlington Colliery, near Radstock, Somerset, England (NMW Specimen 90.8G.8). d. Cluster of *Calamostachys paniculata* Weiss cones, Lower Pennsylvanian Series, Cattybrook Claypit, near Bristol, England (NMW Specimen 86.101G.54a). e. *Macrostachya infundibuliformis* (Brongniart) Schimper, Farrington Formation (Middle Pennsylvanian Series), Kilmersdon Colliery, near Radstock, Somerset, England (NMW Specimen 90.9G.4). From Cleal & Thomas (1999).

Box 6.4 | Identifying *calamostachyalean* cones

The cones are relatively commonly found and (unlike lycophyte cones) are relatively open structures so that the internal arrangement of the various parts can often be seen. They have been assigned to several morphogenera, including *Calamostachys*, *Palaeostachya* and *Macrostachya* (Fig. 6.14 c–e). All of the cones have alternating whorls of sterile bracts and sporangiophores with each usually having two or four sporangia. The different genera are mainly distinguished by the position of the sporangiophores relative to the bracts (Fig. 6.15). Most of these cones were homosporous, but *Calamocarpon* had sporangia containing just a single megaspore, although this can be difficult to recognise as an adpression. Differentiating morphospecies of these has tended to be based on the size and shape of the cones, and the shape of the bracts. However, it is now evident that the type of spores borne by the cones is also an important taxonomic character, and it is often helpful if the type of foliage the parent plant produced is known.

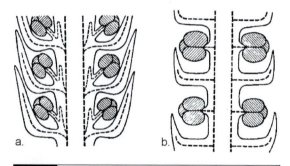

Fig. 6.15 Schematic longitudinal sections through calamite cones showing the relative positions of the sporangium-bearing sporangiophores to the sporophylls on the cone axes. Dashed lines represent the vascular tissue. a. *Palaeostachya*, where the sporangiophores are attached in the axils of the sporophylls. b. *Calamostachys*, where the sporangiophores are attached mid way between sporangiophore whorls. After Boureau (1971).

Gondwana sphenophytes

Other sphenophytes were present in temperate and high palaeolatitudes during Late Palaeozoic times. A number of different kinds have been found in the Lower Permian rocks of Gondwana, notably in India, Australia and South Africa. Many superficially resembled the whorls of leaves borne on the Carboniferous calamites from the palaeotropical belt, but their whorled leaves were fused basally forming a saucer-like sheath around the stem and their reproductive structures are unknown. Another Gondwana horsetail was

Phyllotheca, which was a modest-sized plant and more on the scale of modern *Equisetum* than the giant *Calamites* of the contemporary palaeotropical floras (Fig. 6.16). Also as in the modern *Equisetum*, the leaves were fused at their base to form a clear sheath around the stem. However, the cones had more in common with the Calamostachyaceae in having alternating whorls of bracts and sporangiophores, the latter albeit being more complex in branching twice to produce four peltate heads, each with four sporangia.

Phyllotheca probably grew as dense, single-species stands around the margins of shallow lakes that have been interpreted as analogous to modern-day pastures in today's temperate latitudes. As such they were probably a major food-source for some of the large vertebrate animals that were starting to appear at this time, such as the mammal-like reptiles, the dicynodonts.

Modern sphenophytes

The modern Equisetaceae differ from the Calamostachyaceae in having bractless cones, stems with no secondary growth and leaves that are fused at the base to form prominent leaf-sheaths. Fossils that correspond to the modern type of horsetail are known from Tertiary macrofloras and are confidently assigned to *Equisetum*. However, there are also fossils in Mesozoic and the latest Palaeozoic macrofloras

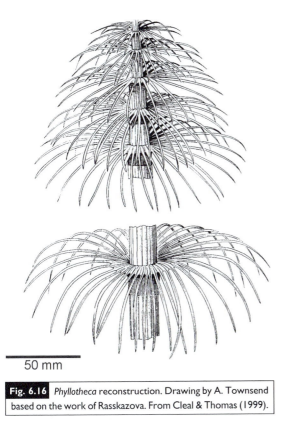

50 mm

Fig. 6.16 *Phyllotheca* reconstruction. Drawing by A. Townsend based on the work of Rasskazova. From Cleal & Thomas (1999).

that resemble *Equisetum* but which are not well enough known or show subtle differences from the living forms, and these are called *Equisetites*. They had developed leaf-sheaths, as in *Equisetum*, but the stems were uncharacteristically large, suggesting that there may have been some secondary growth. They are also sometimes found associated with cones in which there were occasional whorls of sterile bracts separating zones of sporangiophores. These factors have led some authors to question whether the Calamostachyaceae should be regarded as a separate family. It should also be remembered that some of the Palaeozoic leaf whorls seem to show a narrow collar of fusion at the base of the leaves, although this is nowhere near as prominent as the leaf sheath of *Equisetum* and *Equisetites*. This may simply reflect the fact that the groups of taxa that we call families are artificial constructs, and that the fossil record is bound to reveal intermediates between what at first sight might seem like sharply delineated groups.

The origin of the modern Equisetaceae has been the subject of considerable debate, with different palaeobotanists arguing for the Archaeocalamitaceae or the Calamostachyaceae as its immediate ancestor. The absence of bracts in the archaeocalamite cones has, in particular, been used to argue that they were ancestors of the modern family. However, emphasis has recently been given to the presence or absence of elaters on the spores, which, it is thought, are unlikely to have evolved independently more than once. As they are present in the Calamostachyaceae and Equisetaceae, but absent in the Archaeocalamitaceae, it seems most likely that the Calamostachyaceae represent the direct ancestors of modern sphenophytes.

Recommended reading

Abbott (1958), Bashforth & Zodrow (2007), Bateman (1991), Batenburg (1977, 1981), Bureau (1964, 1971), Cleal (in Benton, 1993), Gastaldo (1981, 1992), Good (1971, 1975), Mamay & Bateman (1991), Rayner (1992), Thomas & Cleal (1993).

Chapter 7

Ferns

Ferns are the most conspicuous, diverse and widely distributed group of spore-bearing plants, as well as being the largest and most diverse group of vascular plants after the angiosperms. They are characterised by having large leaves (fronds) that evolved from planated and laminated branching systems, in contrast to the stem outgrowths (microphylls) of the lycophytes and sphenophytes. The great diversity in growth form and life histories has enabled the group to be a successful component of most types of terrestrial plant communities, and occasionally to become the dominant plants.

There are a number of different growth forms of ferns (Fig. 7.2). Some may appear as a cluster of leaves coming from a small erect stem; others as isolated leaves coming from a creeping rhizome that enables them to spread vegetatively over considerable areas. A few ferns reproduce vegetatively by means of bulbils on their leaves or buds on their roots. Some even grow into tall erect tree ferns (Fig. 7.1).

The majority of the ferns are homosporous with their sporangia usually aggregated into clusters, called sori, on the underside of the ultimate segments (pinnules) of divided fronds. The sporangia are often distinctive and can be specific to a family. A few ferns are heterosporous with specialised capsule-like structures that allow them to survive periods of drought.

Ferns have a complex evolutionary history. Their fossil record reveals several extensive systematic changes through geological time, contrasting with the more gradual changes shown by the lycophytes and sphenophytes (Fig. 7.3).

Fig. 7.1 An undisturbed stand of tree ferns in Hawaii in the nineteenth century. The taller trees are one of four species of *Cibotium* endemic to Hawaii. The trunk of a tree fern contains a large central pith surrounded by a cortex in which there are one or more rings of individual vascular strands. They never secondarily thicken their conducting tissues, although a ring of sclerenchyma is often formed on the outside of the stem to give it rigidity. Large leaves are produced by the apical meristem and, as the fern grows taller, the old leaf bases remain to become enclosed in a mass of roots that give the trunk a fibrous appearance. Growth rates vary, with some species increasing by as much as 0.5 m a year in their early years. Individual specimens of some slow growing, forest tree ferns such as *Dicksonia antarctica* Labill. may last for over 200 years.

Fig. 7.2 Living ferns with a range of growth forms and habitats. a. *Polystichum setiferum* (Forssk.) Woyn. growing in a woodland near Cardiff, Wales. A number of fronds arise from a short stem. b. *Polypodium* sp. growing near Cardiff, Wales. Individual fronds arising from a creeping rhizome. c. *Pteridium aquilina* (L.) Kuhn (bracken) covering an open habitat near Caerphilly, Wales. Extensive underground rhizomes produce upright stems from which fronds grow; the stems and their fronds die back in winter. d. *Cheilanthes vellea* (R.Br.) F.Muell. in a lava field in Tenerife, Canary Islands. The woolly felt on the underside of the leaves cuts down water loss through transpiration. e. *Lygodium* sp., a climbing fern found in the Kakum Forest, Ghana. Photos by P. Russell (a–c) and B. A. Thomas (d–e).

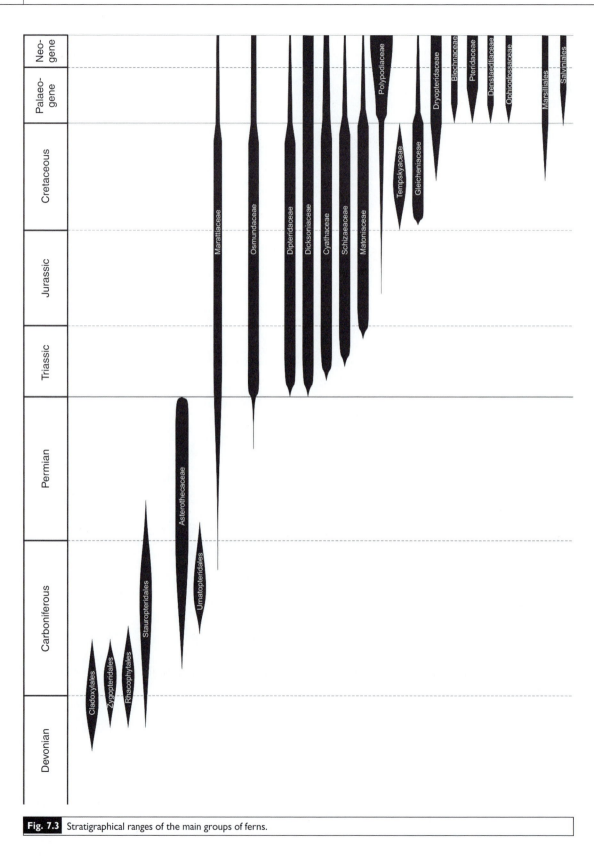

Fig. 7.3 Stratigraphical ranges of the main groups of ferns.

The earliest Devonian ferns were replaced in Mississippian times by different families which themselves went into decline during the Permian Period. Some families of extant ferns had their beginnings in Late Palaeozoic times although the most diverse of our extant families evolved and diversified from Cretaceous times onwards.

The first ferns

The first fern-like plants appeared towards the end of Early Devonian times, originating as a group of plants with new structural and reproductive characters. There is a general acceptance that they arose from within the trimerophyte complex (see Chapter 4). However, there is still debate about whether ferns are a monophyletic group or whether the early ferns (Cladoxylales, Zygopteridales and/ or the Stauropteridales) arose separately from different trimerophyte ancestors. At first, all Devonian plants with large dissected leaves were thought to be ancestral ferns until it was discovered that many were in fact progymnosperms (see Chapter 4).

Many of these early ferns are only known from small fragments and there are the usual problems of interpreting the morphology of the whole plant. Nevertheless, we do know that by Middle Devonian times, two groups of ferns had evolved, the Cladoxylales and the Rhacophytales, and that they retained their identity through to Mississippian times. Both were prominent members of these early plant communities and fossilised remains of the Rhacophytales form the bulk of the earliest coals. The Cladoxylales had a characteristic branching pattern, while a semi-arborescent habit has been proposed for the genera *Pseudosporochnus* and *Calamophyton*. They had a dissected or very deeply lobed xylem, which often gave off strands into the stem appendages, sometimes originating from different lobes of the stele. The small dichotomising ultimate appendages are interpreted as leaves (Fig. 7.5).

The two other groups of early ferns, the Zygopteridales and the Stauropteridales, appeared in late Devonian times although neither formed dominant components of the vegetation. Most of the Zygopteridales have small solid vascular strands, although some have larger ones with internal parenchyma. Within this group is a tendency towards aggregation of the sporangia into sori on the underside of reduced pinnules. The Mississippian *Symplocopteris* from Queensland, Australia is the oldest zygopterid fern showing clear anatomical and morphological distinction between stems, leaves and roots. It is also the oldest known example of a fern with an erect false trunk composed of intertwined branching systems and leaf bases embedded in adventitious roots (Fig. 7.6a). The group diversified during Mississippian times (Fig. 7.6b), with the Stauropteridales (Fig. 7.6c) being the first ferns to be heterosporous.

These early ferns diversified and spread during Carboniferous times and by the Permian Period there were forms such as the central European *Asterochlaena* that showed a combination of stellar and petiole characters of earlier forms. There was also the small tree fern *Dernbachia* from north-east Brazil that had a large dissected stele and advanced dorsiventral petioles which suggest it is somehow related to *Asterochlaena* and to the other Late Palaeozoic fern *Ankyropteris* now included in the Tedeleaceae (see Filicalean Ferns section).

By the end of Palaeozoic times, all four groups of early ferns had become extinct, but by then they had been replaced by many other types of more advanced ferns that showed a much greater diversity of forms.

Modern ferns

Modern ferns are usually split into two main groups on sporangial characters. In one group, consisting of the Marattiales, Urnatopteridales and the Ophioglossales, the sporangia develop from several cells. Mature sporangia are relatively large, thick-walled and contain large numbers of spores (sometimes over 1000); these are called eusporangia (Fig. 7.8d). The other, much larger group of filicalean ferns has smaller, thin walled sporangia that develop

---

Box 7.1 | Describing parts of fern fronds

The most commonly found macrofossil remains of ferns are parts of their fronds. Although fern fronds can vary in form, by far the most common are pinnate fronds. The main parts of a pinnate frond are shown in Fig. 7.4. There is a basal stipe (stalk) that bears the laminate blade. The blade may be undivided, or divided by two or more orders of lateral branching. Most of the segments produced by this lateral branching are known as pinnae (singular pinna), each of which consists of a rachis bearing laterally attached subsidiary segments. The last order segment (i.e. the segment which is not divided into subsidiary segments) is often referred to as a pinnule. Within any one frond, the degree to which it is divided can vary: in the proximal (lower) part of the frond it might be tripinnate, but in the distal (upper) part this will change progressively to bipinnate and then once-pinnate, until at the end there is an undivided segment (or terminal pinnule).

The differentiation of pinnules from pinnae is usually self-evident, although there are levels within the frond where pinnae grade into pinnules. Such intermediate forms are referred to as pinnatifid pinnules. In most ferns, pinnatifid pinnules make up only a small part of the frond, although there are exceptions such as those of the Late Carboniferous *Lobatopteris*. There can also sometimes be problems if the pinnules are themselves lobed. A lobed pinnule can usually be distinguished from pinnae because the lobing is dichotomous, in contrast to the lateral pinnae lobing, but the distinction is not always clear-cut.

Botanists and palaeobotanists often refer to the pinnate divisions of the frond in different ways. This is because the botanist usually deals with whole fronds, whereas the palaeobotanist tends only to have broken fragments of the frond. The botanist will refer to the segments attached to the main rachis as primary pinnae; the primary pinna then consists of a secondary rachis that bears secondary pinnae, and so on. In contrast, palaeobotanists usually call the rachis that bears the pinnules the ultimate rachis, and the ultimate rachis together with the pinnules is known as the ultimate pinna. Ultimate rachises may be attached to penultimate rachises, which, together with the ultimate pinnae, are known as the penultimate pinnae.

---

from only one cell, and contain relatively few (often 64) spores (e.g. Fig. 7.8c, e, f); these are called leptosporangia.

There are a number of families of extant taxa that have long fossil histories and there are a few families consisting entirely of extinct taxa. Some of the extant families had their origins either in Late Palaeozoic or Early Mesozoic times, and their appearance was usually followed by rapid diversification. Certainly by the end of the Palaeozoic Era there were ferns with the form of rhizomatous herbs, epiphytic herbs, shrubs, vines and tree ferns.

Marattiales

The Carboniferous marattialeans could be up to 10 m tall with a 1 m wide trunk, and were the first modern group of ferns to be thought of as real tree ferns. Their stems, when found with anatomy preserved, are referred to the genus *Psaronius* (Figs. 7.9, 7.10a). The fronds expanded from crosiers at the apex of the stem (Fig. 7.11e) into much divided leaves, which may have been up to 3 m or more in length. When the frond eventually died and was shed from the plant, it left a characteristic scar on the stem (Fig. 7.7). The remains of marattialean fronds are widespread in Pennsylvanian and Permian macrofloras, a few examples of which are shown in Fig. 7.11. *Cyathocarpus* fronds were large, up to 2–3 m long, but with small, tooth-shaped pinnules that had a simple venation. The sori usually consisted of four relatively large, elongate sporangia attached to the underside of the pinnules in two rows on either side of the midvein. As with living marattialeans, the sporangia dehisced longitudinally along a zone of thin-walled cells, to release their numerous spores. The spores were small, about 20 μm in diameter, trilete, and ornamented with minute lumps (punctae). *Lobatopteris* is another genus of Pennsylvanian ferns, which differs from *Cyathocarpus* in having larger, more elongate and often deeply lobed (pinnatifid) pinnules. The fructifications were very similar to those of *Cyathocarpus* and there is no doubt that

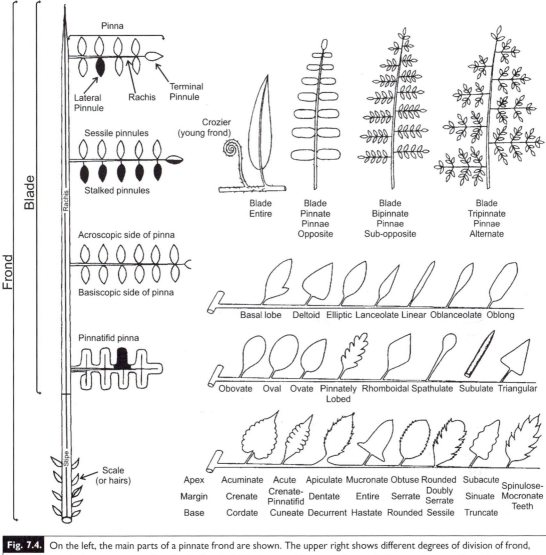

Lobatopteris was also a marattialean. The distinctive pinnatifid pinnules occurred abundantly as adpressions in the Pennsylvanian and different species occur at different stratigraphical levels, making them a useful type of fossil for dating these strata.

The Marattiales was the first modern group of ferns to become a major part of the vegetation. At first these tree ferns were common but not abundant members of the Late Carboniferous flora, being mainly restricted to drier habitats such as the raised river levees. Towards the end of the Carboniferous Period, however, they had replaced the arborescent lycophytes as the dominant trees in the forest swamps, at least in Europe and North America. Many of the coals found in these higher strata are the remains of peat produced by these ferns. The remains of *Cyathocarpus* occur abundantly in the uppermost Middle Pennsylvanian and Upper

Fig. 7.5 Fertile shoot of *Calamophyton bicephalum* Leclerq & Andrews from the Eifelian Stage, Belgium. Scale bar = 10 mm. These dichotomising branches bear sterile and fertile branches in low helices (Palaeobotany Department, University of Liège, Specimen ULg 5011/609B). Photo by M. Fairon-Demaret.

Pennsylvanian series of Europe and North America, where it was probably one of the main peat-producing plants of the time. In contrast *Lobatopteris* never became a dominant part of the back swamp vegetation, always keeping to the drier habitats.

Based on the organisation of sporangia into synangia, it is generally accepted that these late Palaeozoic plants belonged to the Marattiales (Fig. 7.14). However, on the Palaeozoic fronds, sporangia generally had radial synangia, whereas modern marattialeans (except *Christensenia*) have bilateral synangia. Even though

Palaeozoic-type marattialeans continued into Triassic times, there is a lack of transitional forms between the Palaeozoic and Mesozoic forms, which has led some palaeobotanists to put the Palaeozoic genera into their own family, the Asterothecaceae.

The most likely candidates for the ancestral forms of modern marattialeans are the Permian genera *Quasimia* from Saudi Arabia and *Dizeugotheca* from Patagonia. Other genera followed and the marattialeans continued to diversify. The group reached its climax in numbers and diversity in Jurassic times, with such genera

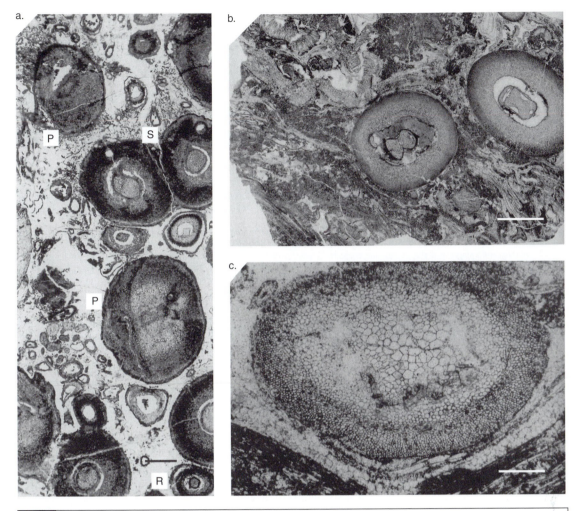

Fig. 7.6 a. *Symplocopteris wyatti* Hueber & Galtier. Transverse section of the central part of the trunk showing several stems (S), phyllophores, (P) and roots (R) occupying the space between stems and phyllophores. Scale bar = 2 mm. b. Transverse section through two petioles of the zygopterid fern *Metaclepsidropsis duplex* from the Lower Carboniferous of Scotland. The petioles are parts of the frond of a scrambling plant and have hour-glass shaped vascular traces. The petiole on the right has a curved trace leading to a pinna. The petiole on the left shows a leaf trace divided into two. NMW 11.64G.2. Scale bar = 2 mm. c. Transverse section through the stem of *Stauropteris oldhamia* from the Lower Pennsylvanian of northern England. It has the four-lobed vascular strand characteristic of the genus. NMW 98.24G.4. Scale bar = 0.5 mm. Photos by J. Galtier (a; see Huber and Galtier, 2002 for more information) and from Cleal & Thomas (1999) (b, c).

as *Marattiopsis* (Fig. 7.15) and *Danaeopteris*. Some fossils are so much like extant taxa that they have been referred to the genera of living ferns, *Marattia*, *Danaea* and *Angiopteris*. The family then steadily declined in number and prominence and there are only a small number of tropical species alive today. Although some species still have very large fronds, none have trunks.

Other Late Palaeozoic ferns

There were other smaller herbaceous ferns living in the Coal Measures swamps or on the riverside levees. They usually had more incised pinnules than those of the marattialeans and were at first mostly referred to the morphogenus *Sphenopteris* (Fig. 7.13). As with the pecopterids, however,

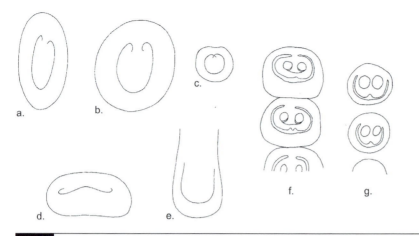

Fig. 7.7 Typical examples of leaf scars on Late Palaeozoic tree fern stems. a. *Caulopteris anglica* Kidston with oval elongated scars (c. 60 mm long) and a single vase-shaped vascular trace. b. *C. cyclostigma* (Lesquereux) Kidston with round scars (> 40 mm long) and a single vase-shaped vascular trace. c. *C. arberi* Crookall with round scars (< 40 mm long) and a single vase-shaped vascular trace. d. *Megaphyton gwynnevahnii* Crookall with leaf scars broader than long (< 60 mm long) and a W-shaped vascular scar. e. *M. frondosum* Artis with leaf scars longer than wide (up to 100 mm long) and an open-vase shaped vascular scar. f. *Aristophyon goldenbergii* (Weiss) Pfefferkorn with oval scars (> 35 mm long) and two vascular scars – the outer with a deep indentation on the lower side; adjacent scars may be contiguous. g. *A. approximatum* (Lindley & Hutton) Pfefferkorn with smaller scars (< 35 mm in diameter) that are usually separated. Drawings by D. M. Spillards. From Cleal & Thomas (1994).

Box 7.2	How to identify Late Palaeozoic marattialean fern stems

Such stems with leaf scars, when preserved as adpressions or casts, are assigned to various form-genera depending on the shape and distribution of the scars. They are identified on size and shape of the scar outline, the number of vascular traces, and the shape of the vascular traces (Fig. 7.7).

improved understanding of these ferns has resulted in the establishment of more natural fossil genera, such as *Bertrandia*, *Corynepteris*, *Crossotheca*, *Hymenophyllites*, *Oligocarpia*, *Renaultia*, *Sphenopteris*, *Urnatopteris* and *Zeilleria*. The affinities of these ferns are still a matter of some conjecture, but they probably belong to various extinct families such as the Botryopteridaceae, Zygopteridaceae, Crossothecaceae and Urnatopteridaceae.

Ophioglossales

This is a small order of homosporous, eusporangiate ferns made up of three genera of extant species: *Ophioglossum*, *Botrychium* and *Helminthostachys*. They are simple in appearance having only one vegetative frond and one fertile frond (sorophore). The cosmopolitan *Ophioglossum* with 25–30 species, has undivided fronds. *Botrychium* is also cosmopolitan with about 50 species and has pinnately dissected fronds. The monotypic *Helminthostachys zelaynica* found in southeast Asia, has a palmate vegetative frond and an undivided fertile frond.

Although the order has been traditionally included with the Marattiales in a loose grouping of eusporangiate ferns and is supposed to be of great antiquity, its origins are as yet completely unknown. For many years, its entire fossil record was based upon rather dubious records of dispersed spores. However, fragments of Palaeocene vegetative and fertile fronds have

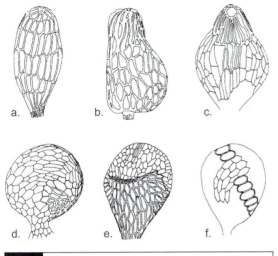

Fig. 7.8 Most fern sporangia have specialised cells enabling them to split open and release their spores. Thickened cells in the annulus shrink when drying thereby setting up differential strains in the other sporangial cells, which rupture to form an opening for spore release. The more complex fern sporangia have a distinct area of thin-walled cells (stomium).
a. Carboniferous *Botryopteris* with no distinct annulus.
b. Carboniferous *Senftenbergia* with a terminal multilayered annulus. c. Extant *Schizaea* with a terminal single-layered annulus. d. Extant *Osmunda* with a lateral patch of annular cells.
e. Extant *Gleichenia* with a lateral annulus. f. Jurassic *Aspidistes* with a linear annulus and a stomium. Redrawn from Thomas & Spicer (1987).

since been found in Alberta, Canada, and these establish an indisputable, although still fairly short, fossil record for the Ophioglossales. These fossils are similar in structure to the extant *Botrychium virginianum* which interestingly grows in Alberta today (Fig. 7.16). They have one pinnately dissected vegetative frond and one pinnately dissected fertile spike. The fact that these very modern-looking fossils are found in very early Neogene times suggests that the order had earlier origins, so we await further discoveries that may throw light on their evolutionary history.

Filicalean ferns

Leptosporangiate ferns originated in Carboniferous times and recent evidence from fertile filicalean frond segments from eastern North America suggests that they may even have evolved near the start of the Carboniferous Period. However, a note of caution must be added before accounts are given of the types of filicalean ferns found in the fossil record. The classification of these ferns is based on a combination of reproductive and anatomical features, so simple matching on overall appearance is certainly not sufficient for accurate interpretation of fossils. Fossils must show characters which are unique to any taxon before they can be included within it. If this is possible, the fossils can then be included within extant families, genera or even species.

One of the best-known examples of misleading overall appearance is the Carboniferous leptosporangiate fern *Senftenbergia plumosa*, whose fronds occur abundantly in the Pennsylvanian Coal Measures of Europe (Fig. 7.11d). It was a tree fern, superficially very similar to the Carboniferous marattialeans dealt with above, with large fronds bearing small, dentate pinnules. However, the fertile structures were quite different from those of the marattialeans. The fertile pinnules bore sporangia, singly or in loose clusters, near the margin and associated with veins. They have a clear apical annulus (Fig. 7.8b), which at one stage was regarded as evidence that these ferns belonged to the extant family, the Schizeaceae. However, there are a number of other features of the plant as a whole, such as the presence of small, vascularised scales on the stem, which are not found in the Schizeaceae. Furthermore, *Ankyropteris brongniartii*, an anatomically preserved fern that is closely related to, if not conspecific with, *Senftenbergia*, has been shown to have axillary branching, a feature that is very unusual in ferns and is more normally associated with seed-plants. The consensus now seems to be that *Senftenbergia plumosa* is a primitive leptosporangiate fern, belonging to an extinct Palaeozoic family, the Tedeleaceae.

Osmundaceae

Many modern groups of ferns appeared after the Permian/Triassic Extinction Event (see Chapter 11

Fig. 7.9 a. Reconstruction of the Carboniferous tree fern with a *Psaronius* trunk. The conical shape of the plant is formed by masses of adventitious roots that help to support the stem. b. Reconstruction of a cut portion of a *Psaronius* stem without the surrounding mantle of roots, showing the concentric rings of conductive strands. The outer ring gives off the conducting tissue to the fronds. The oval areas on the outer surface are leaf scars with vascular traces. c–g. Sections through the trunk, upwards, showing the increasing size and complexity of the vascular system as the enlarging apical meristem produces more tissues to the growing stem. By P. Dean (a–b), or redrawn from Thomas & Spicer (1987; c–g).

for more details of this event and its effect on plant life). The Osmundaceae, which is regarded as the most primitive family of the filicalean ferns, appeared in Late Permian times in both the northern and southern hemispheres. It has the most extensive fossil record of any family of ferns, containing more than 150 extinct species with about half known as adpressions, about 50 from permineralised stems, and 20 from isolated spores.

Today there are three genera: *Osmunda* with about ten species (Fig. 7.17), the southern hemisphere *Leptopteris* with about six species, and the monotypic *Todea barbara* that can sometimes develop into small tree ferns. One distinctive character of living taxa is that each species has two different and quite distinctive types of frond: vegetative fronds in which the pinnules have a broad lamina, and fertile fronds in which the sporangia-bearing segments have very little if any lamina.

Much of the geological history of the Osmundaceae is based upon permineralised stems. They are characteristically trunks, within which the stem is surrounded by a mantle of leaf bases and roots (Fig. 7.10b). Such stems have many circular or oval vascular strands embedded in parenchyma and the leaf petiole bases have an adaxially curved vascular strand. A number of characters have been used to distinguish and name osmundaceous stems, but the arrangement of the thickened cells (sclerenchyma) in the leaf bases seems to be the most reliable.

The structural diversity of the fern axes, and the widespread distribution of fossil and living

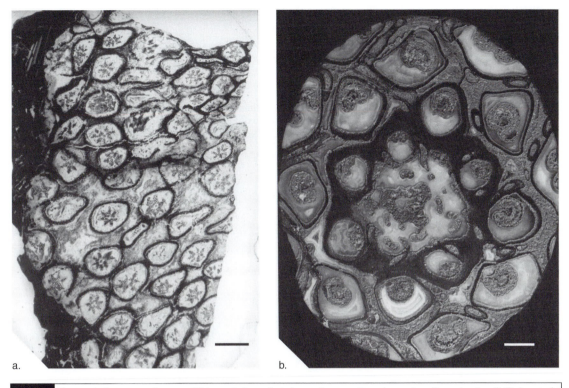

Fig. 7.10 Anatomically preserved stems of Palaeozoic ferns. Scale bars = 5 mm. a. *Psaronius* sp., Late Pennsylvanian silicified peat of Autun, France (NMW Specimen 11.108.84). Photo by NMW Photography Department. b. *Palaeosmunda williamsii* Gould, Permian, Coleron, near Rockhampton, Queensland, Australia (BMNH Specimen V.5000). From Cleal & Thomas (1999).

genera and species, indicate that the group originated in Early Permian or possibly even earlier times. *Grammatopteris* from Germany, France and Brazil and *Rastropteris* from Taiwan are both of early Permian age, while *Palaeosmunda* is known from Late Permian floras of Queensland, Australia and Russia.

Evolution within the Osmundaceae was rapid during Late Palaeozoic and Early Mesozoic times, with many new genera and species found in the fossil record. There was the Middle Triassic *Donwelliacaulis* from Arizona, and Triassic specimens from Antarctica that are indistinguishable from the extant species *Osmunda claytonensis*. Then, in middle Mesozoic floras, other fossil stems from British Columbia in Canada, Utah in the USA, Queensland and Tasmania in Australia, and South Africa are referable to *Osmunda* and recognisable at a subgeneric and even specific level. However, even though their rhizomes

and petioles can be closely compared with those of living members of the Osmundaceae, it is often much more difficult to be so certain about foliage identification unless based on fertile specimens. Nevertheless, some leaves have been assigned to living genera where they are nearly identical to the foliage of living taxa. The presence of fossils of such fertile foliage shows us that the family had developed its habit of dimorphic fronds prior to, or during, Jurassic times. *Todites* and *Cladophlebis* are both common Mesozoic members of the Osmundaceae.

By Palaeocene to Eocene times, the extant *Osmunda* was a characteristic genus of the Arcto-Tertiary flora (see Palaeogene and Neogene Periods in Chapter 11) with records from North America, Greenland, Europe, China and Japan (Fig. 7.17c). *Leptopteris* has a few doubtful records of Miocene petioles from Queensland, Australia while *Todea* has no known fossil record.

Fig. 7.11 Pennsylvanian pecopteroid ferns. a. *Cyathocarpus arborescens* (Brongniart) Weiss. Scale bar = 10 mm. From the Pennsylvanian Subsystem of Chaumoune Canton, Switzerland (BMNH Specimen WC.223 and GC.789). b. *Lobatopteris miltoni* (Artis) Wagner. Scale bar = 10 mm. From Middle Pennsylvanian strata of Clay Cross, Derbyshire (British Geological Survey Specimen 5073). c. Close-up of the same specimen coated with ammonium hydroxide to enhance the surface detail, showing pinnules with two rows of sporangial clusters. Scale bar = 5 mm. d. *Senftenbergia plumosa* (Artis) Stur. Scale bar = 5 mm. From Middle Pennsylvanian strata at Warren House Opencast near Wakefield, Yorkshire (NMW Specimen 92.20G.1). e. Crosier of a Middle Pennsylvanian marattialean fern from Sandwell Park, near Dudley, West Midlands (BMNH Specimen V.1265). From Cleal & Thomas (1999).

Box 7.3 | How to identify Late Palaeozoic pecopteroid ferns

The Palaeozoic marattialeans are known as both petrifactions and adpressions; a fact that has at times led to a dual system of names. The large fronds disintegrated either prior to abscission or fossilisation, so they are only preserved in a fragmentary way. The fronds of the marattialean and tedelacean tree ferns mostly had unlobed pinnules, or elongate linear pinnules with lobed lateral margins, and when preserved as adpressions used to be assigned to the morphogenus *Pecopteris*. However, as more has become known about these fossils, better-defined fossil genera have been developed, based on a combination of the sporangia attached to the pinnules, the form (especially the amount of lobing they show) and the vein pattern of the pinnules. *Pecopteris* is still used for those species of fronds that are less well-known, especially when details of the reproductive structures are unknown.

Most Palaeozoic marattialean ferns are characterised by the sporangia being grouped into clusters (sori), with a row of sori on either side of the midvein of fertile pinnules; there are usually four to six sporangia per sorus. Four genera are usually recognised (Fig. 7.12). The earliest-occurring was *Lobatopteris*, in which the fronds had a large number of lobate pinnules, in which the veins usually forked more than once and in a distinctive pattern. *Acitheca* had similar-sized pinnules to *Lobatopteris*, but which were only rarely lobed, the veins of the pinnules forked in a different way, and the sporangia were much more elongate. *Cyathocarpus* had much smaller pinnules than the other two morphogenera, and the veins tended to be simple or only once-forked. *Ptychocarpus* also usually has simple or just once forked veins, but the veins tend to bend away from the base of the pinnule.

In contrast to the marattialean ferns, the foliage of those ferns included in the Tedeleaceae is characterised by the sporangia being borne singly or in loose clusters, near the pinnule margin and associated with veins. The pinnules can be very variable within a frond, from small and dentate, to elongate and slender, to subtriangular. The veins are usually once forked at a wide angle, or simple. These fronds preserved as adpressions are referred to the fossil genus *Seftenbergia*.

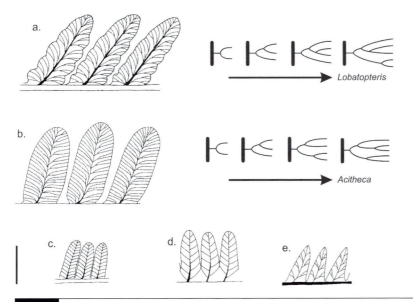

Fig. 7.12 Pennsylvanian pecopteroid pinnules. Scale bar = 5 mm (refers only to drawing of pinnules, not the schematic vein diagrams). a. *Lobatopteris miltoni* (Artis) Wagner, with a high proportion of pinnatifid pinnules, and the third fork of the veins occurring in the middle branch of each vein-cluster. b. *Acitheca polymorpha* (Brongniart) Schimper with tongue-shaped pinnules, often with a slightly constricted base, and the third fork of the veins occurring in the lower branch of each cluster. c. *Cyathocarpus hemitelioides* (Brongniart) Mosbrugger, with smaller, tongue-shaped pinnules and simple veins. d. *Ptychocarpus unitus* (Brongniart) Zeiller, with pinnules that are often decurrent, and veins that curve towards the distal part of the pinnule. e. *Senftenbergia plumosa* (Artis) Stur, with subtriangular pinnules and veins that mostly fork once at a wide angle. Pinnules drawn by D. Spillards, schematic venation diagrams by C. J. Cleal.

Box 7.4 | How to identify Late Palaeozoic sphenopteroid ferns

These pinnules are characterised by being lobed or digitate (Fig. 7.13). Identification is based on a combination of a number of characters that refer them to genera and to species. The problem is that generic characters are those which the majority of fragmentary specimens do not show, such as the overall morphological architecture of the frond, epidermal characters and the type of reproductive organs. Many species even show variation of pinnule characters within one leaf, which is again something that cannot be seen in small fragments. Therefore, in identifying fragments the best way to proceed is to refer the fossils to a species and from this identification the generic name can be assigned. The important characters used for distinguishing species are: the overall form of the pinnules including shape, extent of lobing, outline and degree of fusion to the rachis; the type of venation including the density of veins, the number of times they divide and at what angle they meet the margin.

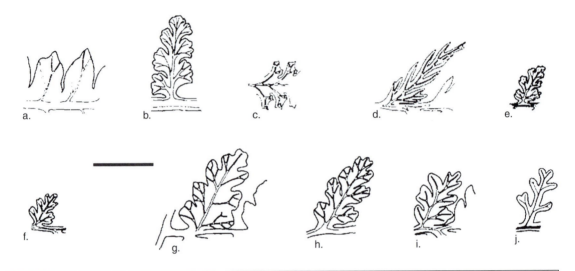

Fig. 7.13 Small 'sphenopteroid' Carboniferous ferns. Scale bar = 5 mm. a. *Corynepteris similis* (Sternberg) Kidston with sub-triangular shallowly lobed pinnules fused at their bases at c. 90° to the rachis. b. *C. coralloides* (Gutbier) Bertrand with sub-cuneiform pinnules with rounded lobes at c. 90° to the rachis and a broad rachis. c. *C. angustissima* (Sternberg) Němejc with small, more or less parallel-sided pinnules fused at their bases and with angular lobes. d. *Sphenopteris sewardii* Kidston with pinnules at < 45° to the rachis and pinnule lobes with acuminate apices. f. *Hymenophyllites quadridactylites* (Gutbier) Kidston with bluntly lobed pinnules that are not decurrent at their bases and a slender rachis (< 1 mm wide). The sporangia, when present, are restricted to the pinnule margin. f. *Sturia amoena* (Stur) Němejc with pinnules at > 45° to the rachis, the pinnules have lobes about as long as broad and gradually taper to an unlobed apex. g. *Sphenopteris rutaefolia* Gutbier with pinnules at > 45° to the rachis. The pinnules have broad lobes that are usually > 1 mm wide, and up to 5 veins may reach the margin of each lobe h. *S. selbyensis* Kidston with angular lobed pinnules at > 45° to the rachis. i. *S. schwerinii* (Stur) Zeiller with pinnules at >45° to the rachis, slender lobes on the pinnules, usually < 1 mm wide, usually 3 or less veins reach the margin of each lobe. j. *S. souichii* Zeiller with deeply incised pinnules < 3 mm long, decurrent at base and ultimate rachis < 1 mm wide. Drawings by D. M. Spillards. From Cleal & Thomas (1994).

Schizaeaceae

The family today consists of four tropical or subtropical genera, and about 160 species, with most being creepers or climbers (Fig. 7.1e). Their large sporangia are borne singly and have a very primitive type of dehiscence with a terminal annulus (Fig. 7.18). *Lygodium* is a pantropical genus of about 40 species that scrambles and climbs by means of leaves that are of indeterminate growth. Its fertile leaves have cone-like structures (sorophores) protruding from the edge of

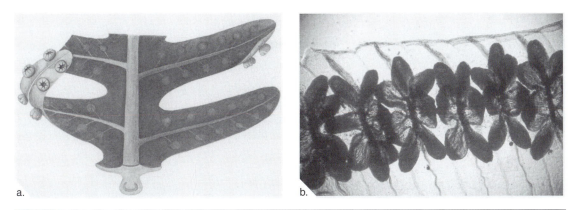

Fig. 7.14 Marattialean Ferns. a. Reconstruction of a fertile pinna of a Carboniferous marattialean fern with pecopteroid pinnules. There are two rows of radial synangia on each pinnule. Painting by P. Dean. b. The edge of a cleared frond of living *Angiopteris evecta* (G. Forst.) Hoffm. showing nearly marginal sori, with six or seven large sporangia, positioned over the veins. Photo by B. A. Thomas.

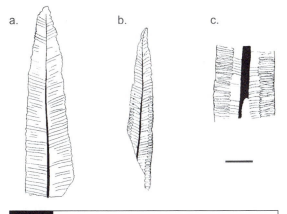

Fig. 7.15 *Marattiopsis hoerensis* (Schimper) Thomas (Marattiaceae) from the Triassic – Jurassic boundary interval of Greenland. Scale bar = 10 mm. a. Upper part of a sterile pinna. b. Upper part of a fertile pinna. c. Median part of a fertile pinna. From Harris (1931).

their lamina. Each lobe of the sorophore is interpreted as a reduced pinna containing a single sporangium.

The family's origin seems to have been in Triassic times. Early members include the genus *Klukia*, which was common in Late Triassic, Jurassic and Cretaceous floras, and *Paralygodium*, which occurred in the Late Cretaceous floras of Japan and an Eocene flora of western Canada. There are also fossils that can be assigned to extant genera, such as *Anemia* from Lower Cretaceous rocks and *Lygodium* from Upper Cretaceous rocks, in Wyoming and New Jersey, USA.

There are many records of post-Mesozoic schizaeaceous ferns with many of them referable to the extant genus *Lygodium*. Some are fertile and closely resemble extant species, like the excellently preserved and widespread North American Eocene *Lygodium kaulfussi* that is very similar to the living *L. palmatum* of eastern North American (Fig. 7.19a–b). Others referable to *Lygodium* are known from Europe, North America, Chile, Australia, New Zealand, and possibly eastern Russia and China. However, the only conclusive evidence of macrofossils of the extant genus *Schizaea* is from European Quaternary floras.

Gleicheniaceae

Members of the Gleicheniaceae grow enormous branching leaves through a system of apical dichotomies and often form dense and impenetrable barriers on the edge of tropical forests (Fig. 7.20).

The earliest known records of gleicheniaceous ferns are from Lower Cretaceous rocks of England and Upper Cretaceous rocks of Japan and New Jersey (USA). Most are named as species of *Oligocarpia* and *Gleichenites* (e.g. Fig. 7.19c) and only a very few are doubtfully referred to the extant *Gleichenia*. The post-Mesozoic macrofossil record is very limited, with the family known only from an Oligocene flora of Tasmania, Eocene floras of western Canada and southern England, and Miocene and Pleistocene floras of

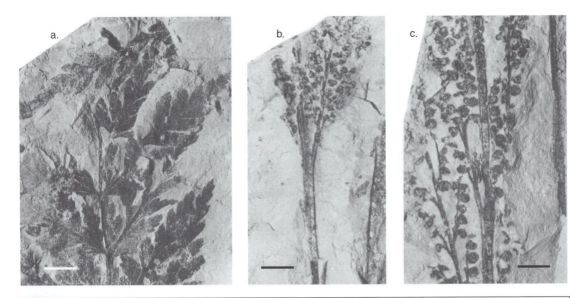

Fig. 7.16 *Botrychium wightonii* Rothwell & Stockey, from the Palaeocene of Alberta, Canada. Scale bars = 5 mm. a. Portion of the sterile frond. b. Fertile frond with an elongated basal region. c. Apical part of fertile frond with sub-opposite divergence of pinnae and alternate rows of sporangia. From Rothwell & Stockey (1989).

Australia. This reduction in distribution might be due to competition from more successful flowering plants. Dispersed spores assignable to the family are recorded from Cretaceous times onwards, and suggest a more widespread distribution of the family in Palaeogene and Neogene times. However, there are several problems in using such records because the spores may have been reworked from older sediments and they may not belong exclusively to the Gleicheniaceae. It therefore remains to be proven that this type of spore found in pre-Palaeogene ferns definitely belongs to the family.

Matoniaceae

Members of the family Matoniaceae first appeared in Late Triassic times in the western USA. These fossils, belonging to the genus *Phlebopteris*, are both abundant and well preserved, showing the plants to have had palmate leaves with an odd number (5–15) of spreading pinnatifid pinnae (Figs. 7.19d, 7.21b). Reconstructions suggest that the leaves grew from horizontally growing rhizomes. Their sori are in single rows on either side of the pinnule midrib and contain 14–20 sporangia each with an oblique annulus. All these characters clearly suggest affinity with living members of the Matoniaceae. However, because the *Phlebopteris* fossils have so many matoniaceous characters, it is likely that plants with at least some of the characters existed earlier. By Jurassic times, matoniaceous ferns were widespread around the world and they continued to be so in Cretaceous times with such successful genera as *Weichselia* forming dense swards by rhizomatous growth, rather like modern day bracken (Fig. 7.20a). Even so, climatic change in middle Cretaceous times in the northern hemisphere brought havoc to the family, which by Late Cretaceous times seems to have died out throughout most of the world, probably again having been out-competed by flowering plants. However, although there are no known matoniacean fossils of Late Cretaceous or later age, the family survived, but became restricted to only two living genera, *Matonia* and *Phaperosorus*, found exclusively in the Malaysian-Borneo area.

Dipteridaceae

Like the Matoniaceae, this family is today very restricted in distribution. There is only one extant genus, *Dipteris*, with eight species

Fig. 7.17 a, b. *Osmunda regalis* L., commonly known as the royal fern growing in South Wales. a. Whole plant; b. Close-up of fertile fronds. c. Sporangium of *Osmunda lignitum* (Giebel) Stur, from the Eocene Series of Germany, showing it split open along the stomium. Scale bar = 0.1 mm. Photos by P. Russell (a–b); drawing (c) by M. Barthel.

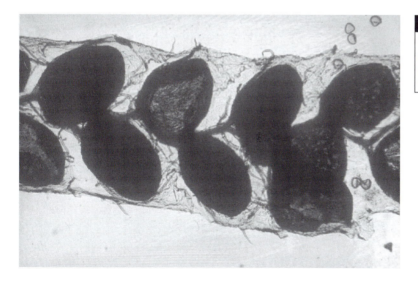

Fig. 7.18 Part of a cleared fertile frond of *Lygodium* sp. (Schizaeaceae) showing large sporangia. Taken from specimen shown in Fig. 7.1e. Photo by B. A. Thomas.

that are only found in the Indo-Malaysian – Polynesian region. The leaf architecture is quite distinctive in the way that it develops through a succession of unequal dichotomies resulting in a flattened, almost umbrella-like frond (Fig. 7.20d).

The members of the Dipteridaceae have a very similar pattern of distribution in time and

Fig. 7.19 a, b. *Lygodium kaulfussi* Heer, from the Eocene Series of Wyoming, USA (Florida Museum of Natural History, Gainesville, Specimens 5307 (a), 5304 (b)). a. Sterile leaf. Scale bar = 10 mm. b. Non-laminar fertile branch. Scale bar = 5 mm. c. *Gleichenites pulchella* Knowlton, from the Cretaceous of Wyoming, USA showing the characteristic dichotomising branching pattern of the leaf. d. *Phlebopteris smithii* (Daughtery) Arnold (Matoniaceae), from the Petrified Forest National Park, New Mexico, USA. Scale bar = 10 mm. Photos by S. Manchester (a, b) and S. R. Ash (d), or taken from Knowlton (1913) (c).

Fig. 7.20 *Dicranopteris* (Gleicheniaceae) in a clearing in a pine forest in Belize. Photo by B. A. Thomas.

Fig. 7.21 Reconstructions of Mesozoic ferns. a. *Weichselia* (Matoniaceae). b. *Phlebopteris smithii* (Matoniaceae). c. *Clathropteris* (Dipteridaceae). d. *Hausmannia* (Dipteridaceae). Paintings by P. Dean (a) and A. Townsend (b–d). From Thomas & Cleal (1998).

space to those of the Matoniaceae suggesting a close relationship between the two families. The origin of the Dipteridaceae appears to have been in Early Triassic or Late Palaeozoic times and, like the Matoniaceae, had a worldwide distribution during Mesozoic times but then became very rare in Late Cretaceous times. There are no records so far of macrofossils from post-Mesozoic floras and the spores are not diagnostic enough for use as stratigraphical markers.

Dicksoniaceae and Cyathaceae

The Dicksoniaceae contains seven extant genera with about 46 species of mainly southern hemisphere, tropical and subtropical tree ferns. An exception is *Culcita*, which includes terrestrial and epiphytic ferns with massive prostrate trunks. *Culcita macrocarpa* is unusual in growing in the Iberian peninsular and the Azores, Madeira and the Canary Islands. The most diagnostic character of species of this family is their

Fig. 7.22 *Cyathea princeps* Mayer (Cyatheaceae) in Xalapa, Mexico. The trunk can grow to 15 m in height and are often covered with brown scales. The fronds can be up to 5 m in length. Photo by B. A. Thomas.

marginal to submarginal sori that are protected by a bivalved structure, the valve nearest the midvein consisting of a cup-shaped membrane (known as an indusium) and the more marginal valve formed by the reflexed margin of the leaf lamina.

The Cyatheaceae contains four extant genera with about 500 species of terrestrial tree ferns with a pan-tropical and temperate southern hemisphere distribution (e.g. Fig. 7.22). One species, *Cyathea gracilis*, is sometimes epiphytic. The most diagnostic character of species of this family is their elongated or rounded sori enclosed by a thin indusium.

The fossil records of both families extend from Early Triassic times, with species referable to the Dicksoniaceae appearing first. Permineralised stems are most common at this time, but in Jurassic and Early Cretaceous floras leaves are the commonest remains. Dicksoniaceous ferns were common in the Jurassic Period, belonging to one of several genera such as *Coniopteris* (Fig. 7.23), *Kylikipteris*, *Erobacia* (Fig. 7.24) and *Dicksonia*. *Coniopteris* fronds, when fertile, have sori enclosed in cup-shaped indusia on the ends of reduced pinnule segments. The Dicksoniaceae became gradually restricted to the southern hemisphere during Late Cretaceous times, and only a very few definite fossil representatives of later ages have been found in the northern hemisphere.

Fertile fronds referable to *Cyatheae* are known from calcareous concretions of Early Cretaceous age from British Columbia, Canada, demonstrating that essentially modern species of cyatheaceous tree ferns had evolved by this time.

Fig. 7.23 *Coniopteris* from the Spitzbergen Palaeogene. From Kvaček & Manum (1993).

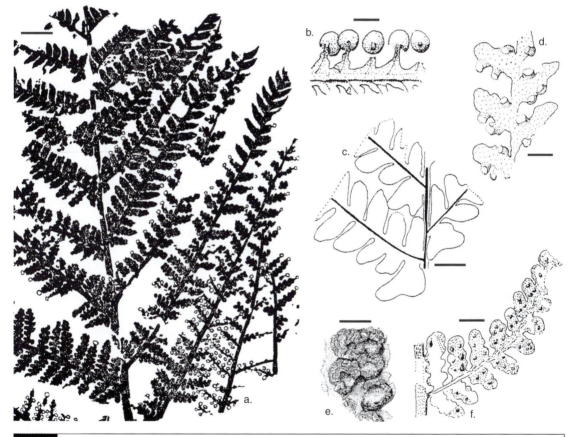

Fig. 7.24 Middle Jurassic ferns from Yorkshire, England. a–b. *Kylikipteris arguta* (Lindley & Hutton) Harris. a. Terminal portion of a frond. Scale bar = 10 mm. b. Enlargement to show the sori. Scale bar = 2 mm. c–d. *Eboracia lobifolia* (Phillips) Thomas. c. Part of sterile frond. Scale bar = 5 mm. d. Fertile pinnules showing sori. Scale bar = 2 mm. e–f. *Aspidistes thomasii* Harris. e. Close-up of fertile pinnule showing sori. Scale bar = 1 mm. f. Part of fertile pinna. Scale bar = 2 mm. From Harris (1946) (a, b) and Wilson & Yates (1953) (c–f).

Polypodiaceous ferns

The family Polypodiaceae is arguably the most abundant and widespread fern family today (Fig. 7.25). However, the family has been constantly changed in content and often divided into many, much more narrowly defined, families such as the Blechnaceae, Dryopteridaceae,

a.   b.

Fig. 7.25 *Dryopteris affinis (Lowe) Fraser-Jenkins growing in woodland near Welsh St Donats, Wales. a. Complete plant. b. Close-up of underside of fertile pinnules showing sori, each covered by a kidney-shaped indusium. Photos by P. Russell.*

Thelypteridaceae, Hymenophyllaceae, Grammitidiaceae, Lophosoriaceae, Dennstaedtiaceae, Pteridaceae, Ophioglossaceae and the redefined Polypodiaceae. The only characters that are constant throughout the Polypodiaceae in its broad sense relate to the sporangia: they are small, usually stalked, have a vertical, incomplete annulus and contain 64 spores. In some genera, the sporangia are covered by a protective structure called the indusium.

Several Mesozoic fossils have been referred to the broad Polypodiaceae and included in genera of extant ferns such as *Aspidium*, *Asplenium*, *Davallia*, *Polypodium* and *Pteris*. However, the major problem in evaluating and accepting these claims is that nearly all of them are sterile foliage, so in fact very few can be referred to these genera, or even to a polypodiaceous family, with any certainty. The earliest example that might possibly be acceptable as this type of fern is *Aspidistes thomasii* (Fig. 7.24e–f) from the Yorkshire Jurassic, but even here there is no unequivocal proof.

There are a number of post-Mesozoic fossil ferns that are definitely polypodiaceous, although there are others which are only known from portions of sterile foliage and therefore very debatable in affinity. A few examples are given here of the better known fossil ferns, but this is not the place for a full account of them all. It is also most likely that these ferns are under-represented in the fossil record, because today most of them have fronds that die, collapse and wilt on the plants in contrast to the flowering plants and conifers that shed their leaves in vast numbers. Because of this, post-Mesozoic leaf floras, in particular, give an over-representation of flowering plants and an under-representation of ferns.

Fragmentary specimens from Cretaceous and later rocks have been assigned to the extant genus *Onoclea* (Dryopteridaceae), which is regarded as belonging to the most advanced group of filicalean ferns, the dennstaedtioid-asplenioid line. The best published example of a fossil belonging to this genus has even been referred to the living species *Onoclea sensibilis* that now grows throughout northern Asia and central and eastern North America. Thousands of specimens of this species were recovered from Palaeocene non-marine flood deposits in Alberta, Canada, including large segments of whole plants in positions of growth, within a flora dominated by a birch relative (*Paleocarpinus*) and the Dawn Redwood (*Metasequoia*). Today, *Onoclea sensibilis* grows in moist woodlands and along the sides of streams and marshy lakes. Other anatomically preserved rhizomes and fern bases called *Wessiea* from the middle Miocene floras of Washington State, USA, conform to the *Onoclea*-type of anatomy.

b.

a.

b.

Fig. 7.26 a. *Woodwardia radicans* (L.) Sm. in Madeira. Photo by M. Rickard. b. *Woodwardia virginica* (L.) J. E. Smith from the middle Miocene Series of Washington State, USA. Scale bar = 2 mm. Photo by R. A. Stockey.

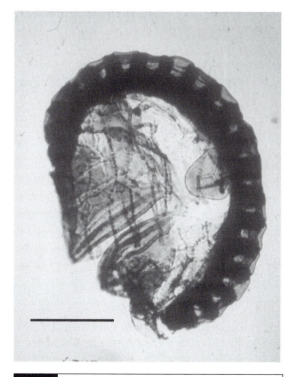

Fig. 7.27 Dispersed sporangium of *Acrosticum anglicum* Collinson (Palaeogene deposits, Isle of Wight, England) showing a vertical annulus and an open stomium. A trilete spore can be seen in the sporangium. Scale bar = 1 mm. Photo by M. E. Collinson.

Several species of *Woodwardia* (Blechnaceae) have been described from various Tertiary circum-Arctic floras in the USA, Canada, Europe, Kamcatchka and Japan. Anatomically preserved rhizomes, vegetative pinnae and fertile pinnae of *Woodwardia* are also known from middle Miocene floras of Washington State, USA (Fig. 7.26), and there are also specimens referable to *Osmunda* in the same matrix. This Miocene flora demonstrates a fern assemblage similar to those of some temperate floras of today. *Woodwardia* now grows naturally in North America, Europe and Asia.

Other undisputed fossil ferns are much less spectacular, although they do show recognisable sporangia. For example, *Acrostichum* (Pterideaceae) has three extant species living in mangroves with a total tropical to warm temperate distribution. They have distinctive large pinnae with reticulate venation, which helps to identify fragments in the fossil record. In post-Mesozoic times, it was a widespread genus in Europe, North America, the Ukraine, Siberia, Manchuria and Japan. Many have the distinctive sporangia and dispersed sporangial masses, while isolated sporangia from the Palaeogene floras of Britain have been referred to *Acrostichum anglicum* (Fig. 7.27).

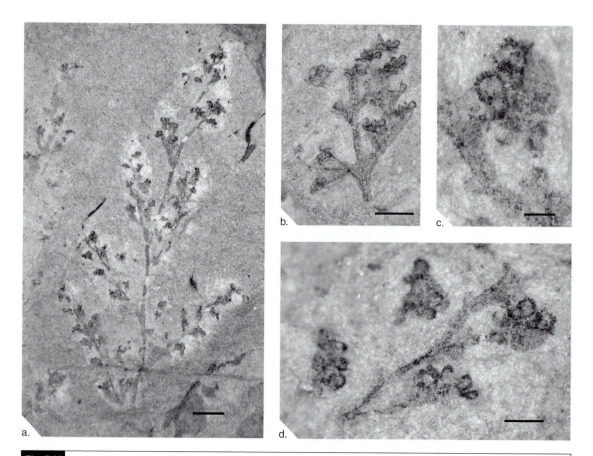

Fig. 7.28 *Hopetedia praetermissa* Axsmith *et al.* A filmy fern from the middle Perkins Formation (Late Triassic age), North Carolina, USA (see Axsmith *et al.*, 2001, for further details). a. Pinna. Scale bar = 2 mm. Complete frond with sori at the tips of the distal lobes. b. Close-up of fertile pinnules showing sporangia. Scale bar = 0.5 mm. c. Sorus with sporangia showing cells of annulus. Scale bar = 0.1 mm. d. Close-up showing several small pinnules. Scale bar = 1 mm. Photos provided by T. N. Taylor and reproduced with permission.

From the available evidence, it appears that these were once associated with lakes and freshwater marshes.

Very delicate ferns are sometimes preserved in the fossil record. For example, there is the filmy fern *Hopetedia praetermissa* from Triassic rocks of North Carolina, USA (Fig. 7.28) that appears to have a filmy texture. Its indusium has a funnel-like structure similar to those in the extant filmy fern *Trichomanes*.

Occasionally, anatomically preserved rhizomes can be closely compared with those of extant ferns allowing them to be identified with reasonable certainty: for example, rhizomes of *Dennstaedtiopsis* (Dennstaedtiaceae) have been described from Oregon, USA, and several others belonging to the family from the Eocene London Clay of south east England and similar-aged deposits in Belgium.

Tempskya

Although most Mesozoic fern fossils can be accommodated within the modern fern families, one rather bizarre form is quite different from anything alive today. This is the Cretaceous tree fern *Temskya*, which grew throughout North

America, Europe and Japan. There is one possible Palaeogene record from southern England, although its age is very debatable because it was found in a loose pebble. The plants grew upwards forming a false trunk of several inter-twining stems and roots (Fig. 7.29). All the leaves were small and spread over the upper portions of the trunks unlike the large leaves of other tree ferns that grow in an apical crown. We only know of this fern through petrified portions of the false trunk, and associated leaf bases and roots, and nothing is known of its reproductive structures. However, its mode of growth is so different from that of any other known fern, either alive or extinct, that there is general agree-ment that it must belong to its own family, the Tempskyaceae.

Heterosporous ferns

There are two orders of extant heterosporous ferns, the Marsiliales and the Salviniales. Both produce their megaspores and microspores in sporangia that are encased in desiccation-resistant structures called sporocarps that are attached by short stalks at the base of fronds. This common feature indicates that both orders probably had their origins in ancestors that grew in seasonal pools.

The Marsiliales contains three genera of rhizomatous rooting ferns that often grow in dense mats. *Marsilea* is quadrifoliate, rather like a tall four-leafed clover in appearance, and has an almost worldwide distribution, except for cool temperate regions, and contains about 64 species. *Pilularia* has rounded grass-like leaves, and has five species with a cosmopoli-tan distribution. The monospecific *Regnelidium* is bifoliate and is restricted to Brazil and Argentina.

Many of the earlier records of fossil *Marsilea* have been shown to be incorrect with some of them even being petals of angiosperm flowers. There are reliable specimens from mid Creta-ceous rocks of Kansas, USA, together with asso-ciated structures that might well be sporocarps.

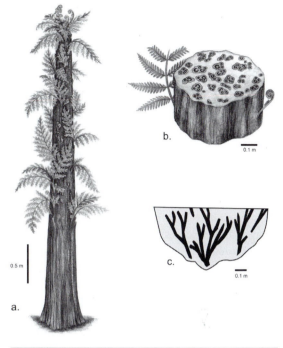
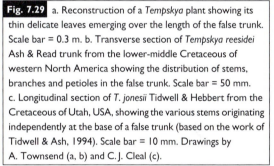

Fig. 7.29 a. Reconstruction of a *Tempskya* plant showing its thin delicate leaves emerging over the length of the false trunk. Scale bar = 0.3 m. b. Transverse section of *Tempskya reesidei* Ash & Read trunk from the lower-middle Cretaceous of western North America showing the distribution of stems, branches and petioles in the false trunk. Scale bar = 50 mm. c. Longitudinal section of *T. jonesii* Tidwell & Hebbert from the Cretaceous of Utah, USA, showing the various stems originating independently at the base of a false trunk (based on the work of Tidwell & Ash, 1994). Scale bar = 10 mm. Drawings by A. Townsend (a, b) and C. J. Cleal (c).

These are *Marsilea johnhallii* (Fig. 7.30a). There are also a few megaspores known from the Creta-ceous Period onwards.

Members of the Salviniales are mostly trop-ical and the family contains two genera: *Salvinia* with about 14 species, and *Azolla* with about 7 species. Both are small and free floating on the surface of lakes and very slow moving streams (Fig. 7.31), and can propagate easily by growth and division. *Salvinia* has surface floating leaves and submerged dissected leaves that produce the sporocarps. *Azolla* has horizontal leafy stems with relatively small overlapping leaves, and sporocarps formed on the underside of the stem. *Azolla* can fix nitrogen and for that reason is commonly grown in rice paddies prior to planting the rice.

The oldest reported record of *Salvinia* and *Azolla* is from near the Cretaceous/Tertiary boundary in India. Other records of *Salvinia* are from near the Palaeocene/Eocene boundary of southern England, and the late Miocene deposits of Poland (Fig. 7.32d). Dispersed megaspores and clumps of microsporangia each held together by a common perispore, called massulae, are widespread from Eocene floras onwards and this has resulted in a double taxonomy based on leaves (*Salvinia*) and megaspores (*Minerisporites*).

Palaeocene vegetative remains of *Azolla* plants are known from Alberta, British Columbia, Saskatchewan and South Dakota in North America (Fig. 7.32a). The plants are up to 22.5 mm long with alternate, imbricated leaves along the branched stems. Some of these fossil remains are presumed to have been fertile on the basis of closely associated massulae and megaspores, some with attached massulae. The sporocarps that these spores were in appear to have decayed prior to preservation. These specimens indicate that most of the vegetative characters of modern *Azolla* had evolved by middle Palaeocene times. *Azolla* has a large fossil record based on preserved sporocarps, and spores such as *Ariadnaesporites* and *Capulisporites*. Its megaspores have floats derived from other aborted megaspores that are also impregnated with sporopollenin. The number of floats is an important character in distinguishing fossil species of *Azolla*. Massulae (Fig. 7.32c) are often found attached to the megaspores as they are in extant species. The number of floats attached to the megaspore was originally thought to decrease from an ancestral Late Cretaceous form with 24 to the low numbers of today's species. New evidence shows that there are other spores with three or even fewer floats of Late Cretaceous age. It seems likely that there was a rapid diversification in Late Cretaceous times that resulted in a variety of float numbers, but that only the few-floated ones survived to the present-day.

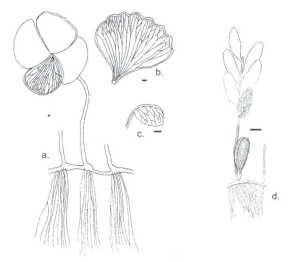

Fig. 7.30 Reconstructions of heterosporous ferns. a–c. *Marsilea johnhallii* Skog & Dilcher, Early Cretaceous Dakota Formation, Kansas, U.S.A. a. The whole plant with the position of three leaves and the roots at the nodes. b. A leaflet. c. The possible sporocarp. Scale bars are 1 mm. d. *Hydropteris pinnata* Rothwell & Stockey from the Late Cretaceous Series of Alberta, Canada, showing the position of two leaves and one sporocarp on the rooting rhizome. Scale bar = 5 mm.

There is one other heterosporous fern, of Late Cretaceous age, from Alberta, Canada that is called *Hydropteris pinnata* (Fig. 7.30b). This fern has rhizomes with pinnate fronds and sporocarps situated at the bases of the leaves. Unbranched roots occur along the length of the rhizome, rather like those in *Regnellidium*. The megaspores have floats and the microspores are clumped into massulae. *Hydropteris* is, therefore, unlike any other heterosporous fern in combining the overall appearance and morphology of the Marsiliales with the reproductive characters of the Salviniales.

These aquatic ferns appear to have undergone a burst of evolution during Cretaceous times that was followed by a general stasis in morphology, reproductive strategy and habitat choice. The remains of both *Azolla* and *Salvinia* are found in freshwater deposits where they are associated with remains of aquatic flowering plants, showing that they occupied similar habitats to those species that live today.

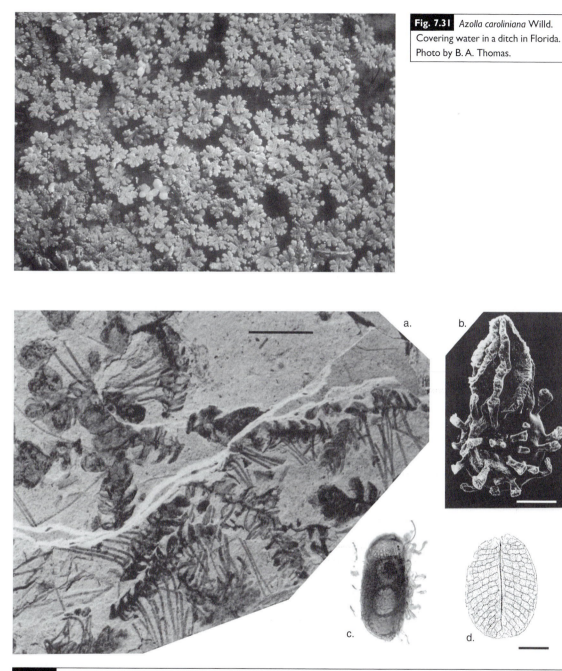

Fig. 7.31 *Azolla caroliniana* Willd. Covering water in a ditch in Florida. Photo by B. A. Thomas.

Fig. 7.32 a. *Azolla primaeva* Dawson, middle Eocene Series, Princeton, British Columbia, Canada. Scale bar = 5 mm. Several lengths of floating heterosporous fern with long roots. Photo by M. E. Collinson. b. *Arcellites hexapartitus* (Dijkstra) Potter, Wessex Formation (Barremian), Isle of Wight, England. Scale bar = 0.5 mm. A megaspore from an extinct group of water-ferns related to the Marsileaceae or Salviniaceae. Photo by D. J. Batten. c. An *Azolla* massula from the Palaeogene Series of Spitzbergen. The massula contains four microspores with a trilete mark visible on the second from the bottom. From Kvaček & Manum (1993), with permission. d. Drawing of a detached floating leaf of *Salvinia mildeana* Göppert from the late Miocene of Poland. Scale bar = 5 mm. From Collinson et al. (2001), with permission.

Recommended reading

Andrews (1952), Axsmith *et al.* (2001), Batten *et al.* (1994), Brousmiche (1983), Camus (1991), Camus *et al.* (1991, 1996), Cleal (in Benton, 1993), Collinson (1980, 2001, 2002), Collinson *et al.* (2001), Dyer & Page (1985), Harris (1961b), Hueber & Galtier (2002), Knowlton (1917), Kvaček & Manum (1993), Jordan *et al.* (1996), Large & Braggins (2001), Manchester & Zavada (1987), Millay (1997), Miller (1971), Pigg & Rothwell (2001), Read & Brown (1937), Rössler & Galtier (2002a, 2002b), Rothwell (1996), Rothwell & Stockey (1989, 1991, 1994), Skog & Dilcher (1992), Thomas & Cleal (1993, 1998), Tidwell & Ash (1994), Zodrow *et al.* (2006).

Chapter 8

Early gymnosperms

The plants dealt with so far in this book had many features that adapted them to life on dry land, notably in the asexual (sporophyte) phase of their life cycle. However, their sexual (gametophyte) phase remained dependent on moisture for fertilisation to take place, and so these plants could not dissociate themselves completely from damp conditions. This constraint was not fully overcome until the evolution of the ovule during Late Devonian times, which not only provided protection for the vulnerable gametophyte but also opened up a range of alternative modes of dispersal for the plant. It proved to be one of the key evolutionary developments in vegetation history, allowing plants to invade all but the most hostile terrestrial habitats.

Seed-bearing plants are normally divided into two general groups: the angiosperms (or 'flowering plants'), in which the seeds are enclosed by a protective ovary (see Chapter 10); and the gymnosperms whose seeds have no such protective ovary and are often referred to as 'naked'. This chapter discusses the gymnosperms that are mainly characteristic of the Palaeozoic Era – the groups known informally as pteridosperms, glossopterids and cordaites (Fig. 8.1). A more detailed review of these plants can be found in Anderson *et al.* (2007).

What are ovules and seeds?

An ovule is the ultimate expression of heterospory (see Chapter 4) with the female sporangium now containing just one viable megaspore (although some early gymnosperms retained three other, aborted megaspores that formed part of the original spore-tetrad). The megaspore wall is now normally referred to as the embryo sac and the sporangium wall as the nucellus.

A characteristic feature of ovules is that the nucellus is surrounded by one or more layers of protective tissue. The innermost layer is known as an integument. In the earliest ovules (sometimes referred to as pre-ovules) found in the fossil record this protective layer consists of a sheath of sterile axes surrounding the nucellus (where it is referred to as a pre-integument), but in true ovules the integument forms a continuous layer of tissue (Fig. 8.2). There are two general models for how this sheath of axes came about. One idea is that ovules evolved from clusters of sporangium-bearing axes, but that all but one of these axes lost their sporangium and the rest formed the integument. The other view is that ovules evolved from fertile fronds, in which the sporangium became surrounded by part of the frond lamina. Although there is no unequivocal evidence one way or the other, the first of these theories is given some support by the fact that gymnosperms are thought to have evolved from the progymnosperms (see Chapter 4) and these are characterised by having sporangia borne in clusters.

Many ovules also have yet another protective layer called the cupule (Fig. 8.2). Not all gymnosperms have a cupule and in some there may be more than one ovule per cupule. As with the integument, the origin of the cupule has been the subject of some debate, but the fact that many early gymnosperm cupules were multi-ovulate

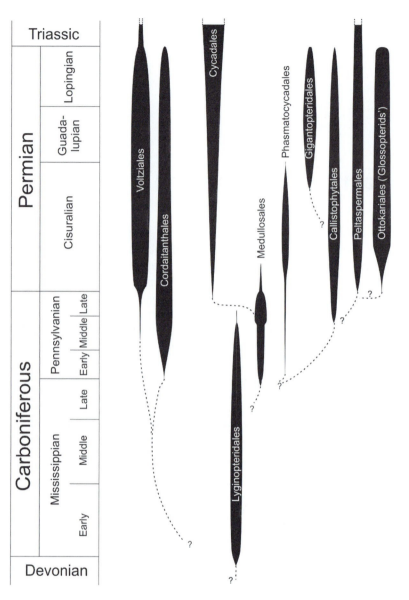

Fig. 8.1 Ranges of the main orders of gymnosperms through the Palaeozoic Era. Based mainly on Anderson *et al.* (2007).

tends to support the idea that they represent vegetative tissue that has surrounded the ovules. It seems likely, therefore, that the integument and cupule evolved in quite independent ways.

In the distal part of the integument in modern-day ovules there is a small hole known as a micropyle through which pollen can enter. In pre-ovules, the distal part of the nucellus was exposed and so obviously there was no micropyle (Fig. 8.4b). Pollen was captured instead on a trumpet-like projection from the nucellus known as a salpinx or lagenostome.

Gymnosperm reproduction

The gymnosperm reproductive cycle is essentially the same as that in early land plants, with alternating sexual and asexual phases (see Chapter 4). However, the sexual phase now takes place entirely within the enclosed environment of the ovule. The megaspore produces the female gametophyte whilst still within the nucellus (sporangium wall). Most living gymnosperms produce a drop of sticky liquid ('pollen drop') which is exuded from the micropyle (or salpinx) to capture

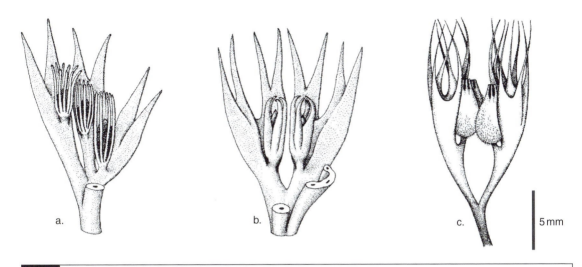

Fig. 8.2 Reconstructions of early ovules showing progressive enclosure of the nucellus within the integument. Note that in all three, there are several ovules enclosed within the cupule. a. *Moresnetia zalesskyi* Stockmans, in which the pre-integument consists of a sheath of axes. b. *Elkinsia polymorpha* Rothwell et al., in which the axes of the pre-integument have started to fuse, but the apex of the nucellus is still visible. c. *Archaeosperma arnoldii* Pettitt & Beck, in which the integument almost completely surrounds the nucellus. Drawing by A. Townsend, based on the work of Pettitt & Beck (c), or taken from Stewart & Rothwell (1993) (a, b).

pollen being carried on the wind. The pollen drop is then drawn back into a cavity below the micropyle known as a pollen chamber. A pollen drop is obviously difficult to preserve in the fossil record, but there is some evidence that they existed in at least some Palaeozoic ovules.

Once inside the pollen chamber, the pollen produces a small male gametophyte. In living cycads and probably the early gymnosperms, the male gametophyte liberates motile gametes that swim through the remaining pollen drop to the archegonia that are now exposed on the top of the female gametophyte. In the more advanced gymnosperms (e.g. conifers) the male gametophyte is itself only a minute structure consisting of just a few cells. The male sperm nuclei travel down a pollen tube that grows out from the pollen grain directly to the archegonia. In some of the early gymnosperms, the pollen retains some features of the ancestral microspores, such as the trilete mark on the surface, but it is assumed that like modern gymnosperm pollen they did not produce free-living gametophytes. Such grains are called pre-pollen.

After fertilisation has taken place, a diploid embryo is produced; an ovule containing a sporophyte embryo now becomes known as a seed. The seed is then released from the parent plant, to grow into a new individual. The seed often has features that help in its dispersal, such as wings for wind-dispersal, or a fleshy outer coat that acts as an attractant for animals to eat it. In the latter case, the seeds are able to pass through the animal's digestive tract unharmed.

In many modern seed-plants (angiosperms as well as gymnosperms), the embryo can remain dormant within the seed and may only germinate when conditions are favourable (e.g. at the end of a dry season or after a period of winter frosts). Among the early fossilised gymnosperms, however, seeds containing embryos are rarely found. Seed dormancy may thus have been a relatively late adaptation among gymnosperms which allowed them to grow in a wider range of habitats.

What plants did gymnosperms evolve from?

Gymnosperms were almost certainly derived from the progymnosperms, because, as well as

Box 8.1 | How to distinguish pteridosperm fronds from fern fronds

This is one of the problems most frequently faced by students starting to study fossil plants but for which there is no easy solution. The obvious point is that most ferns have sporangia on the underside of their pinnules, whereas pteridosperms produced ovules. However, some pteridosperms (e.g. the Callistophytales) produced pollen-organs on the underside of their pinnules, which superficially look very like fern sporangia. Some ferns such as the Osmundales appear not to have their sporangia attached to pinnules, because their sporangia-bearing pinnules have no leaf laminae and therefore look quite different from the purely vegetative pinnules. Then there is the fact that not all fern fronds are fertile, and even in fertile fronds not all pinnules have sporangia.

There are general guidelines for distinguishing the fronds of ferns and seed-plants, but they are not totally reliable, and there are still some species whose systematic position remains uncertain (hence the use of morphotaxa for them, as discussed in Chapter 3).

The presence of sporangia attached to the pinnules usually indicates (with just a few exceptions) that it is a fern.

Ferns tend to have smaller, more delicate pinnules.

Fern pinnules tend to have veins with a more regular branching pattern, whereas in pteridosperms veins tend to branch in a more irregular pattern essentially giving even coverage of veins over the pinnule.

Pteridosperm fronds tend to have a more complex architecture, with pinnae and/or pinnules attached to rachises of different orders. The main rachis also often dichotomises near the base of the frond (although at least one family of extant ferns also shows this feature: the Gleicheniaceae).

exhibiting heterospory, they were the only other earlier plant group that developed extensive dense secondary wood in their stems. Although exact details of the progymnosperm – gymnosperm transition are still unclear, it is likely that the ancestors of the seed-plants were the aneurophytaleans, which had similar large compound leaves (fronds).

The pteridosperms

Many early gymnosperms had large, compound leaves, and during the eighteenth and early nineteenth centuries they were classified as ferns. Later in the nineteenth century, however, it became evident that these plants differed in several ways from true ferns (see Box 8.1) and that they showed anatomical details more

reminiscent of modern-day gymnosperms known as cycads. By the end of the nineteenth century, the general consensus was that these Pennsylvanian plants were a fern-like group showing some gymnosperm-like characters – in essence an evolutionary 'missing-link' between the ferns and the gymnosperms – and they became known as the Cycadofilices. Then, in the early twentieth century, it was discovered that one of these fronds had distinctive glandular hairs on its surface which were identical to hairs borne on a large seed known as *Lagenostoma* that were found in the same rocks. Some other ovules have been reported 'attached' to 'cycadofilicean' leaves, although direct proof that they were actually attached and not just fortuitously preserved lying on the leaf is still lacking. There are also a number of records of clusters of ovules of this type attached to branching axes, which may in turn have been attached to the leaf or maybe to the stem of the plant. Although the exact positioning of the ovules on the plant is still conjectural, it is generally accepted that these large, frond-like leaves were the foliage of primitive gymnosperms, usually referred to as pteridosperms (or sometimes as 'seed-ferns', although this is highly misleading as they really have little to do with true ferns).

Pteridosperms were traditionally regarded as a formal taxonomic group of plants (Pteridosperm-ales or Pteridospermopsida, depending on the rank) implying that they all shared a common ancestor. However, it is difficult to identify a particular character that defines the group other than their large, compound leaves. It was thought that having ovules attached directly to the leaves was a defining character but at least some of the earliest pteridosperms had ovules quite separate from the vegetative fronds, and evidence of attachment in many of the later pteridosperms is at best equivocal. In most other characters, such as stem and ovule anatomy, pollen, and leaf architecture, the group is clearly heterogeneous and it is probable that it does not represent a natural taxonomic group. Nevertheless, the term 'pteridosperm' is a convenient short-hand for describing the fossils of these early gymnosperms, especially the vegetative remains, and it continues to be used in the palaeobotanical literature in this informal sense. Four orders of pteridosperms are usually recognised in Palaeozoic floras: the Lyginopteridales, Medullosales, Callistophytales and Peltaspermales.

Fig. 8.3 *Elkinsia*, one of the earliest known seed plants, of Late Devonian age, from North America. The foliage consisted of highly segmented fronds, while the ovules were borne in clusters at the top of the plant. Drawing by A. Townsend, based on the work of G. W. Rothwell, S. E. Scheckler and W. H. Gillespie.

Lyginopteridales

The earliest known pteridosperms belong to the order Lyginopteridales (Fig. 8.1), and much of our understanding of how and why seeds evolved is based on research on this group of plants. There are two well documented families within the order. The oldest, the Elkinsiaceae, was widespread in late Devonian and early Mississippian times, especially in tropical palaeolatitudes, but was replaced during Pennsylvanian times by the Lyginopteridaceae.

The most completely reconstructed plant of the Elkinsiaceae is *Elkinsia polymorpha* from the Late Devonian Upper Hampshire Formation of West Virginia, USA (Fig. 8.3). It was probably less than 1 m high, with a single upright stem bearing helically arranged, large fronds. The stems had a stele with a trilobed cross section and there was some secondary wood in the thicker stems. The lower part of the *Elkinsia* fronds had a major dichotomous fork of the main rachis, a feature seen in most Palaeozoic pteridosperms. Each branch produced by the fork was further divided by two orders of lateral, pinnate branching, the

Fig. 8.4 Early lyginopteridalean pteridosperms. a. *Sphenopteridium rigidum* (Ludwig) Potonié. Scale bar = 5 mm. Small fragment of a frond probably of the Elkinsiales. Upper Devonian Series, Plaistow Quarry, north Devon, England (BMNH Specimen V.3562). b. *Hydrasperma tenuis* Long. Scale bar = 0.5 mm. Vertical section through an anatomically preserved pre-ovule, showing the pre-integument forming an open collar around the lagenostome and pollen chamber at the top of the nucellus. Middle Mississippian Series, Oxroad Bay, East Lothian, Scotland (Royal Museum of Scotland, Edinburgh, Specimen OBC084hT/48). Photo by R. Bateman. c. *Sphenopteridium pachyrrachis* (Göppert) Schimper showing the main dichotomy of the primary rachis. Scale bar = 10 mm. Middle Mississippian Series, Glencartholm, Scotland (BMNH Specimen V.186). d. *Diplopteridium holdenii* Lele and Walton, showing a stem with numerous fronds attached, at least one of which shows clusters of ovules attached in the fork of the main rachis. Scale bar = 10 mm. Drybrook Sandstone (Middle Mississippian Series), Forest of Dean, England (BMNH Specimen V.62331b). From Cleal & Thomas (1999).

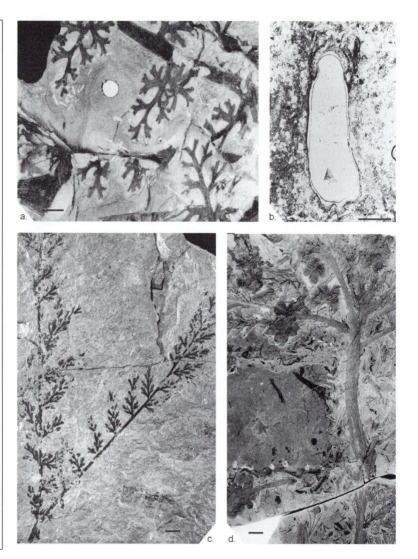

last order bearing narrowly digitate leaflets or pinnules (Fig. 8.4a, b). The vegetative fronds branched in one plane (planated). In contrast, the reproductive structures were borne on three-dimensional branching fertile fronds in which either cupulate pre-ovules or pre-pollen sacs were borne at the ends of the branches (similar to Fig. 8.4d). *Elkinsia* probably had separate male and female fronds. The general form of these fertile fronds has been interpreted as an adaptation to wind fertilisation. On the female fronds, four pre-ovules were borne within each cupule, the latter consisting of sixteen free, elongate segments (Fig. 8.2b).

Fossils of the Lyginopteridales were widespread in Mississippian times, suggesting that the newly evolved ovules gave a significant advantage to the gymnosperms. Many were relatively small plants, like *Elkinsia*, but there were also some substantial sized ones such as *Pitys* with trunks up to 2 m or more in diameter. The foliage was relatively similar with most of the known examples consisting of small fronds with a basal fork of the main rachis, and digitate or lobed pinnules (Fig. 8.4a, c, d). However, the ovules were more diverse in structure, showing a progression from those with an integument consisting of a sheath of separate segments and a

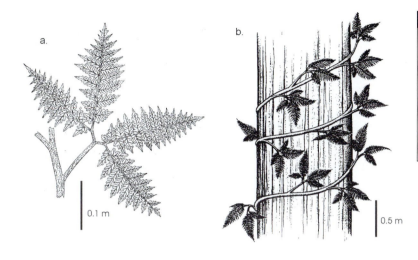

a.

0.1 m

b.

0.5 m

Fig. 8.5 At least some of the Pennsylvanian-aged lagenostomaleans, such as this *Mariopteris*, were probably lianas that grew up trees. a. Reconstruction of *Mariopteris* frond, from Zeiller (1900). b. Reconstruction of a climbing *Mariopteris* plant. Drawings by A. Townsend.

completely exposed lagenostome, to ones where the integumental sheath has become fused but the lagenostome was still exposed, to more modern-looking ovules with the integument entirely encasing the nucellus except for the small distal micropyle (Figs. 8.2, 8.4b). There was also diversity in how the ovules were borne on the plant. In some cases, they were in multi-ovulate cupules, in others the ovules were borne in clusters attached to the angle of the main fork of what appears to be a vegetative frond (Fig. 8.4d).

During Mississippian times, the Elkinsiaceae were progressively replaced by another family of the Lyginopteridales, the Lyginopteridaceae. Like elkinsiacean foliage, Mississippian lyginopteridacean fronds had a major fork in the rachis producing two branches that were more or less mirror-images of each other. The ultimate segments of the fronds were pinnules with digitate lobes. Remains of their stems and rachises often show distinctive surface markings reflecting bands of sclerotic tissue in the cortex: *Lyginopteris* stems, for instance, often show distinctive diamond-shaped surface markings, whereas stems bearing the fronds referable to the other lyginopterid genera *Eusphenopteris* and *Mariopteris* often show transverse bars.

The ovules borne on the later Pennsylvanian-aged lyginopteridalean plants in many ways represent the culmination of the changes that were observed in the Mississippian-aged plants.

The nucellus was now surrounded by an entire integument that had just the small distal micropylar opening to allow entry of the pre-pollen. The nucellus still retained a lagenostome, although it no longer functioned to capture pollen (this strengthens the idea that these Pennsylvanian plants were closely related to the Mississippian-aged lyginopteridaleans). The ovules were borne singly within a cupule, but how these were attached to the plant is uncertain. In some cases, there were clusters of cupulate ovules borne on a truss of slender branching axes but whether these were attached to a frond, or themselves formed separate fertile fronds (as in *Elkinsia*) is not clear.

In Pennsylvanian times, most lyginopteridaleans were relatively small plants. To overcome their lack of height, to obtain light for photosynthesis, which would have been a major problem in forests, many climbed up the trunks of trees as lianas or vines (Fig. 8.5). Examples are particularly well-known from the Pennsylvanian-aged palaeotropical forests, for instance, the plants that had fronds of the fossil genus *Mariopteris* were attached to rope-like stems with an anatomy well adapted to a liana habit. *Mariopteris* fronds were generally similar to those of the Mississippian lyginopteridaleans but they were much smaller (mostly less than 0.5 m long) and their pinnules had more rounded lobes (Fig. 8.6; drawings of other types of lyginopteridalean leaflet are shown in Fig. 8.7).

Fig. 8.6 Pennsylvanian-aged lyginopteridaleans. Both specimens were probably lianas belonging to the fossil genus *Mariopteris*. Scale bars = 10 mm. a. An almost complete frond of *Mariopteris sauveurii* (Brongniart) Frech, showing the proximal dichotomy of the main rachis. A characteristic feature of *Mariopteris* is that the two branches produced by the dichotomy of the main rachis undergo a second non-symmetrical dichotomy, dividing the frond into four segments. Seam 26, Sulzbach Formation (Middle Pennsylvanian Series), Frankenholz Mine, Saarland, Germany (collections of the Saarbrücken Mining School Specimen B/305). Photo by C. J. Cleal. b. Part of a frond of *Mariopteris nervosa* (Brongniart) Zeiller, showing the typical shape and venation of this group of lagenostomaleans. British Coal Measures (Middle Pennsylvanian Series), further details not recorded (NMW Specimen 30.232G.165). From Cleal & Thomas (1999).

Lyginopteridalean diversity and abundance declined significantly during Late Pennsylvanian times and the order eventually became extinct before the start of the Permian Period. We do not know exactly why this was, although it is notable that the decline of the Lyginopteridales coincided with an increase in abundance and diversity of the Callistophytales (a group to be discussed later). The callistophytaleans probably occupied broadly similar habitats to the Lyginopteridales, but had significantly more sophisticated reproductive strategies and may have simply out-competed them.

Medullosales

The earliest known medullosaleans were of Late Mississippian age and they eventually became extinct during Early Permian times (Fig. 8.1). During Pennsylvanian times, however, they were one of the most widespread gymnosperm groups, especially in the palaeotropical floras, and their remains are widespread throughout Europe, North America, Central Asia and China.

The Middle Pennsylvanian Medullosales were mainly small to medium sized trees

Box 8.2 | Identifying pteridosperms

Pteridosperm fronds were complex structures that look superficially similar to fern fronds, and whose main component parts are called the same as those of fern fronds (pinnae, rachises, pinnules, etc., see Box 7.1). The identification of pteridosperm fronds requires an understanding of the morphology of the pinnules and of the way the rachises branch (the latter feature known as the frond architecture). Placing a fossil in a particular species is often based mainly on details of the pinnules: the pinnule size and shape, whether or not the pinnule is constricted at its base, whether or not the pinnule is lobed or digitate, the length of the midvein, and the branching pattern and density of the lateral veins (Figs. 8.7, 8.8). The form of the pinna and the shape of the end (terminal) pinnule can also be a useful guide. This is all fortunate, as most pteridosperm foliage fossils are only fragments of the frond, but are still enough to identify to species.

The generic and higher-rank position is more problematic, however, as it depends on a combination of the frond architecture, features shown by the cuticles, and the anatomy of the rachises. Such characters cannot be distinguished unless there is exceptional preservation of the specimens. In most cases, determining the generic position of pteridosperm fossils depends on identifying it at the species level, and then knowing from earlier studies that that species belongs to a particular genus because someone has reported a rare but exceptionally preserved example showing the diagnostic characters. For instance, most examples of *Laveineopteris rarinervis* (Bunbury) Cleal *et al.* will not show the diagnostic characters of *Laveineopteris*. However, a few rare examples have been reported showing the frond architecture (basal dichotomy of main rachis, intercalated pinna, basal cyclopterid leaves), thinly cutinised lower epidermis, and a rachial anatomy with what appear to be several distinct vascular segments that unequivocally place it in that genus. If it belongs to the *'rarinervis'* species, it has to belong to the genus *Laveineopteris*.

The key architectural features for identifying pteridosperm frond fossil genera are:

Whether there are one or more dichotomies of the main rachis near the base of the frond

Whether or not there are pinnae attached to the main rachis below the basal fork

Whether or not there are pinnules or small pinnae attached ('intercalated') to rachises between pinnae

Whether the pinnae are terminated by one or two pinnules

Whether the surfaces of the rachises show any surface patterning reflecting surface hairs or sclerotic plates in the cortex

Certain distinctive features of the pinnule shape and venation can also be used to separate fossil genera. For instance, *Palmatopteris* is distinguished from *Mariopteris* by the former having deeply incised pinnules whereas the latter have pinnules with rounded lobes or that are in some cases unlobed. Critically, however, their pinnule morphologies are insufficient on their own to define or identify these fossil genera. There are some true ferns that have deeply incised pinnules similar to those of *Palmatopteris*, but these lack the quadripartite fronds and distinctive transverse bars across the rachises.

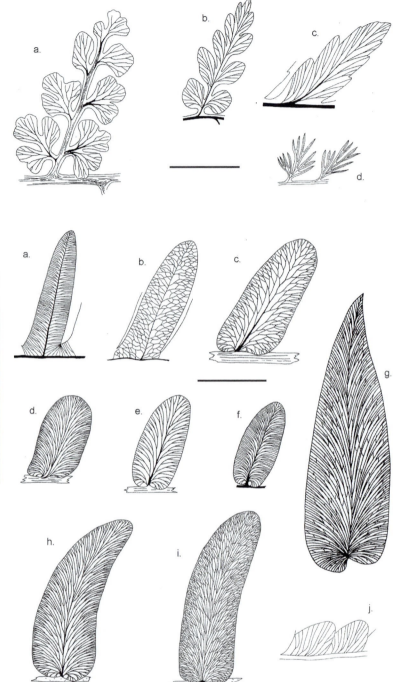

Fig. 8.7 Drawing of typical leaflets of Pennsylvanian Lyginopteridales fronds. Scale bar = 10 mm. a. *Eusphenopteris obtusiloba* (Brongniart) Novik. b. *Karinopteris acuta* (Brongniart) Boersma. c. *Mariopteris muricata* (Brongniart) Zeiller. d. *Palmatopteris geniculata* (Germar & Kaulfuss) Potonié. Drawings by D. M. Spillards. From Cleal & Thomas (1994).

Fig. 8.8 Drawings of leaflets of Pennsylvanian Medullosales fronds. Scale bar = 10 mm. a. *Alethopteris lonchitica* Sternberg. b. *Lonchopteris rugosa* Brongniart. c. *Reticulopteris muensteri* (Eichwald) Gothan. d. *Neuropteris ovata* Hoffmann. e. *Laveineopteris loshii* (Brongniart) Cleal et al. f. *Neuralethopteris jongmansii* Laveine. g. *Macroenropteris scheuchzeri* (Hoffmann) Cleal et al. h. *Paripteris gigantea* (Sternberg) Gothan. i. *Linopteris neuropteroides* (Gutbier) Zeiller. j. *Odontopteris cantabrica* Wagner. Drawing by D. M. Spillards. From Cleal & Thomas (1994).

(Fig. 8.9a), but during Late Pennsylvanian times the order also included smaller scrambling plants. A distinctive feature of the order is that when the stems and rachises are viewed in transverse section, they appear to have contained several quite separate vascular strands separated by softer tissue (cortical parenchyma). A superficially similar vascular system

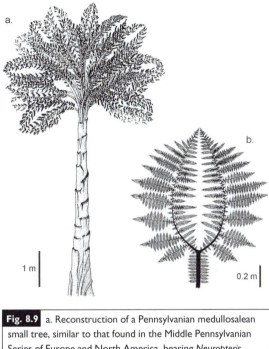

Fig. 8.9 a. Reconstruction of a Pennsylvanian medullosalean small tree, similar to that found in the Middle Pennsylvanian Series of Europe and North America, bearing *Neuropteris* fronds. Drawing by P. Dean. b. Reconstruction of the frond of *Neuropteris heterophylla* Brongniart. Drawing by C. J. Cleal.

occurs in some ferns (e.g. Fig. 7.10) where it is thought to have been the result of a number of separate stems having fused laterally together (resulting in a type of vasculature known as a polystele), but this was not the case in the Medullosales. By carefully mapping the course of these vascular strands along medullosalean stems, they can be seen to have divided and fused repeatedly along their length; each stem and rachis in fact had just a single stele that had become dissected. The resulting vascular structure gave the stem considerable mechanical strength, but without needing to produce large quantities of physiologically 'expensive' secondary wood.

The success of the medullosaleans in the Pennsylvanian palaeotropical forests is reflected in the diversity of foliage preserved in the rocks of this age. The fronds ranged from the 7 m long fronds of *Alethopteris*, which clearly belonged to quite substantial trees, to the much smaller *Odontopteris* fronds that were often only 0.5 m long and probably belonged either to climbing or ground-scrambling plants (part of an average-sized, 2 m long frond is shown in Fig. 8.10a). However, nearly all species had fronds with a basal dichotomy of the main rachis, producing two branches that were three- or four-times divided into fern-like pinnae (Fig. 8.9b).

Most medullosalean fronds had pinnules with simple or dichotomous veins, but others had a meshed or anastomosed venation. This is a simple form of vein-anastomosis, in which the veins became increasingly sinuous until adjacent ones touched and eventually fused. For one genus (*Neuropteris*) it is possible to trace this gradual change from open-dichotomous to anastomosed venation through the sequence of Middle Pennsylvanian rocks in Europe, where it eventually culminated in pinnules with an anastomosed venation and is known as *Reticulopteris*. This simple style of anastomosed veining is quite different from that seen in today's angiosperms, which comprises several different orders of veining.

Medullosalean foliage often had a thick cuticle showing evidence of the underlying epidermal cells (Fig. 8.10b). These cuticles have been the subject of several studies, and the distribution and structure of the stomata have proved important features for helping understand the systematics of these fronds. In many of the fronds, there were numerous small pores known as hydathodes from where the plant could exude (guttate) any excess water. Hydathodes are today found in many plants of the tropical rain forests and are an adaptation to growing in humid conditions.

Most medullosalean ovules and seeds were large and had a characteristic three-fold radial symmetry, often reflected in the presence of three longitudinal ribs (Fig. 8.11). This makes them fairly easy to recognise as adpression fossils, and even internal sandstone casts of the nucellus will show the ribs. They are fairly widespread in the fossil record, but almost invariably detached from the parent plant. We know very little about how they were borne on the plant. There have been recorded examples of individual ovules being found apparently attached to a frond, replacing one of the

Fig. 8.10 *Neuropteris ovata* Hoffmann, a typical medullosalean frond of Middle Pennsylvanian age. a. Large portion of a frond. Scale bar = 0.1 m. Sydney Mines Formation (Middle Pennsylvanian age), Point Aconi, Cape Breton Island, Canada (University of Cape Breton, Specimen 985GF-248). Photo by E. L. Zodrow. b. Cuticle from a pinnule of the same species. The specimen shows the epidermal structure of both the upper (adaxial) surface with strongly sinuous cells, and the lower (abaxial) surface showing stomata and hydathodes. Scale bar = 100 μm. Emery Seam, Sydney Mines Formation (Middle Pennsylvanian age), Glace Bay, Cape Breton Island, Canada (University of Cape Breton, Specimen 985GF-248). Photo by C. J. Cleal.

vegetative pinnules, but it is difficult to be certain that these are not just examples of detached ovules accidentally lying on top of frond fragments. The medullosaleans had some of the largest seeds produced by any gymnosperm, the largest recorded example being 110 mm long. They were almost identical in structure to those of living cycads, suggesting the two groups are probably related. One of their most characteristic features is that the nucellus was always quite separate from the integument, except at the very base of the ovule. In some species there was a rudimentary lagenostome at the distal end of the nucellus but in most it was the micropyle that exuded the pollen drop to capture the pollen.

The pre-pollen grains found in the pollen chambers of most medullosalean seeds resulted from tetrads of grains lying side-by-side in a circle. Each was in contact with two others, an organisation that resulted in a line separating the two contact faces. Such grains with a single

line are said to be monolete. They probably produced motile male gametes, as does the pollen of living cycads. The pre-pollen was produced by compound synangia consisting of clusters of elongate sporangia fused together along their length. These varied considerably in complexity from clusters of just four to over one thousand sporangia. Unlike the ovules, there is good evidence that in at least some medullosaleans the synangia were attached directly to the rachises of fronds, but whether this was the case in all species is not totally certain.

Callistophytales

This was the third order of pteridospermous plants that grew in the Pennsylvanian palaeotropical forests. Although not as abundant as either the Medullosales or Lyginopteridales, they have received considerable attention from palaeobotanists, because, although they look superficially like the more primitive pteridosperms discussed above, they had a more sophisticated reproductive strategy.

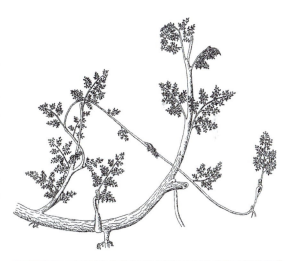

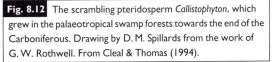

Fig. 8.12 The scrambling pteridosperm *Callistophyton*, which grew in the palaeotropical swamp forests towards the end of the Carboniferous. Drawing by D. M. Spillards from the work of G. W. Rothwell. From Cleal & Thomas (1994).

The best studied (the *Callistophyton*-plant) was a small shrub with a rather scrambling habit (Fig. 8.12). The main stems were up to 30 mm wide, with an anatomy that was rather similar to some lagenostomaleans, with a central pith around which were several strands of primary xylem. This was then entirely surrounded by secondary wood. The cortex around the secondary wood is very distinctive, having numerous cavities lined by secretory tissue that probably produced resin. Similar cavities have been found in associated foliage and reproductive organs, indicating that the various organs belonged to the same plant. The whole plant was reconstructed on the basis of this anatomical similarity and many of these organs have never been found actually connected.

The best known foliage of this group belongs to the fossil genus *Dicksonites*. Like most other pteridosperms, the frond had a near-basal dichotomy of the main rachis. The distinctive pinnules are usually robust and bear a superficial similarity to the pinnules in some Lyginopteridales fronds, such as *Mariopteris*. There is in fact no absolutely foolproof way of distinguishing callisophytalean pinnules from lyginopteridalean pinnules, although the former tend to be somewhat vaulted, whereas the latter are often (but not always) attached to rachises with transverse markings (the sclerotic tissue mentioned previously).

The reproductive organs were borne directly on the underside of pinnules of fronds that look essentially similar to the 'normal' vegetative fronds. Anatomically, the seeds (usually referred to as *Callospermarion*) bear some superficial similarities to those of the Medullosales, having a complex integument that was only fused to the nucellus in its basal part. Unlike the radially symmetrical medullosalean seeds, they appear to be flattened, with two apical horns. However, in the basal part of the *Callospermarion* seed the vascular system had an essentially radial configuration, and it seems likely that this type of seed was in fact derived from an essentially radially symmetrical condition, such as seen in the lagenostomalean seeds.

The pollen-organs are similar to those of other pteridosperms in consisting of radial clusters of sporangia, although in the callistophytes they were attached directly to the underside of pinnules. This gives such pinnules the superficial appearance of fertile fern foliage, although there is no suggestion that they are closely related. Nevertheless, care must be taken when studying fertile callistophytalean fronds, especially when preserved as compressions, that they are not confused with ferns. The most distinctive feature of these pollen-organs is the pollen, which seems to have been true pollen with two bladders presumably to assist in wind dispersal. Such pollen is very similar to that found in cordaites and conifers.

Another plant that grew in the Pennsylvanian palaeotropical forests and that was probably related to the Callistophytales was *Eremopteris* (Fig. 8.13). These plants had much smaller leaves

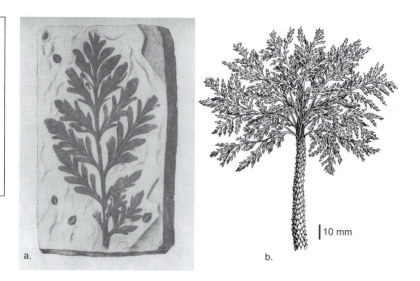

Fig. 8.13 *Eremopteris artemisiaefolia* (Sternberg) Schimper, Middle Pennsylvanian Series, Durham Coalfield, England. a. One of the specimens of leaf figured by Brongniart (1828–38, pl. 47, fig. 12). Even in the early nineteenth century, the regular association of these leaves and the distinctive horned ovules had been noted. b. Reconstruction of the *Eremopteris* plant by A. Townsend.

a.

b.

10 mm

than the other pteridosperms of these forests, usually less than 0.3 m long, and were probably from smaller plants. Although no ovules have ever been found attached, flattened, winged ovules are repeatedly found associated with the *Eremopteris* leaves and it is widely accepted that they belonged to the same plant. Recent work has found that remains of the leaves have small scars on the main rachis and these are of the same size and shape as the attachment-scar at the base (chalaza) of the ovules. It therefore seems likely that this is where the ovules were originally attached. It is because this plant seems to have had flattened ovules attached to the leaves that *Eremopteris* has been suggested to belong to the Callistophytales.

Like the medullosaleans, the callistophytes were very much plants of the Pennsylvanian palaeotropical forests. They first occurred in the middle Pennsylvanian times, reached their zenith in early Late Pennsylvanian times, and in Europe and North America appear to become extinct in Early Permian times. Whether they survived into the Permian palaeotropical forests in China remains to be proved.

Peltasperms

As the medullosaleans and callistophytaleans were going into decline during Permian times, another group of pteridosperms appeared (Fig. 8.1). We have little direct evidence as to the stature of these plants but, from the size of their leaves and the associated stems, they were probably shrubs or small trees. As with most gymnosperms, the foliage is by far the most commonly found remains of these plants. The leaves were small compared to many of the Carboniferous pteridosperms, and could be quite variable in structure. In a few cases, they had undivided leaves, such as *Tatarina* found in the Late Permian temperate floras of western Siberia. Most, however, were fronds similar to those of the Palaeozoic pteridosperms. One of the most typical of the Early Permian fronds in Europe is referred to as *Autunia* (formerly *Callipteris*) and most fronds were only just over 0.5 m long (Fig. 8.14a). Unlike most pteridosperms, this does not have a near-basal dichotomy of the main rachis, although in the apical part the fern-like, pinnate branching sometimes breaks down to produce what are called pseudo-dichotomies. There

Fig. 8.14 Peltasperm foliage. Scale bar = 10 mm. a. *Autunia conferta* (Sternberg) Kerp, showing the characteristic intercalated pinnules on the primary rachis. Lauterecken-Odernheim Formation (Early Permian age), Lebach or Berschweiler, Germany (Weiss Collection, Museum für Naturkunde, Berlin, Specimen 59/9/1). Photo by H. Kerp. b. *Supaia shanxiensis* Wang. Tianlongsi Formation (Early Permian age), Wangtao village, Qinyuan District, Shanxi, China (Specimen 9306–2, Institute of Geology and Mineral Resources, Tianjin, Shanxi, China). Photo by Wang Ziqiang.

was just one order of lateral branching, and pinnules were attached to both these and the main rachis.

At about the same time, peltasperms with another type of leaf, known as *Supaia*, were growing in China and North America (Fig. 8.14b). These were much smaller fronds than *Autunia*, often less than 0.20 m long, and had a major dichotomy of the main rachis. *Supaia* also differs in not having any lateral branching. The form of the leaf is so different from that of *Autunia* that their relationship might be questioned. However, recent work in China has revealed ovuliferous cones associated with *Supaia* leaves which are very similar to those of *Autunia*.

We know very little about the anatomy of the ovules, but they were borne on the underside of disc-like structures, which in turn were arranged in loose cones. The pollen was produced by cup-shaped clusters of several sporangia that were fused at the base.

Although peltasperms first appeared in tropical palaeolatitudes, they extended their range into middle and high palaeolatitude Gondwana and Angara floras during the Permian Period (see Chapter 11). The peltasperms are also known from Triassic floras, so clearly survived the Permian/Triassic extinction event that caused so many of the other Palaeozoic gymnosperms to disappear. However, they eventually disappeared during early Mesozoic times, perhaps being unable to adapt to the heavy grazing by herbaceous dinosaurs that were starting to appear at this time. Alternatively, they may not have disappeared, as such, but simply evolved into one of the better-known Mesozoic gymnosperm groups such as the Umkomasiales mentioned briefly in Chapter 9.

Glossopterids

Permian rocks in India, Australia, southern Africa, South America and Antarctica, often contain abundant remains of leaves (Fig. 8.15) belonging to the general group of plants known as the glossopterids (strictly, they are better referred to as the order Ottokariales, but here we will continue to use the traditional, informal name). The presence of this glossopterid foliage in such widely separated locations and at such different palaeolatitudes was a puzzle to the nineteenth century palaeobotanists, and was not solved until the development of the continental drift model (as discussed in Chapter 2).

Glossopterid leaves were generally entire (Figs. 8.15, 18.16b) and not divided like those of most other pteridosperms, although one cycad-like leaf (*Pteronilssonia*) with glossopterid epidermal structure has been reported from Lower Permian rocks of India. There was considerable variation in leaf-size, the largest entire-margined leaves having reached more than 0.30 m in length. The leaves were almost certainly produced by substantial-sized trees (Fig. 8.15a), although this idea is partly speculative since the leaves are mostly found detached from the stems. The frequency with which detached leaves are found has led some authors to suggest that they were abscised in the autumn, as occurs in most angiosperms living today in temperate climates, but no specialised abscission layer has yet been identified. Wood sometimes associated with the leaves, which looks rather like conifer wood with prominent growth rings, is generally assumed to be the trunk and branches of the tree that bore them.

Several different types of reproductive organ have been found attached to glossopterid leaves or shoots (e.g. Fig. 8.16c), suggesting that this was a very diverse group of plants – perhaps not a totally surprising fact, since they dominated the southern temperate forests for at least 40 million years (i.e. most of the Permian Period). Several ovules were attached to the expanded head of a sporophyll-like structure, which in turn was attached via a slender stalk to a leaf (an example is shown in Fig. 8.15a). Although the intimate association between the foliage and the reproductive organs is not in doubt, there has been some disagreement as to whether these ovulate structures were attached to the midvein of the leaf or to its petiole. The leaves were sometimes indistinguishable from 'normal', purely vegetative leaves, or were

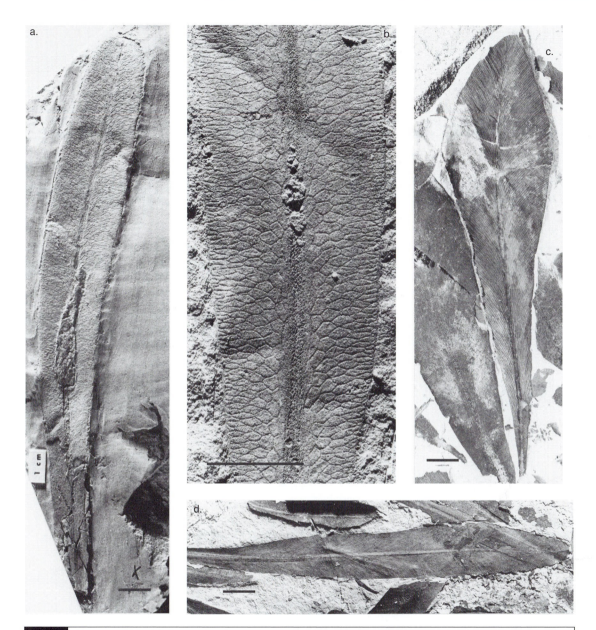

Fig. 8.15 Glossopterid foliage. Scale bars all = 10 mm. a, b. *Lanceolatus lerouxides* Plumstead leaf. Vryheid Formation (Middle Ecca, Early Permian age), Leeukuil Quarries, Vereeniging, South Africa. a. A leaf with an elongate, ovule-bearing structure attached to its midvein (Bernard Price Palaeontological Institute, Johannesburg, Specimen BP/2/-). b. Close-up of a leaf showing the characteristic mesh-veining of this genus (the Vereeniging Museum, Transvaal, Specimen VM/03/3205/22). c. *Lidgettonia mooiriverensis* Anderson & Anderson, showing the non-anastomosed venation characteristic of this genus. Estcourt Formation (Late Permian age), Mooi River, National Road, Natal, South Africa (Bernard Price Institute for Palaeontology, Johannesburg, Specimen BP/2/7563). d. *Lidgettonia lidgettonioides* (Lacey *et al.*) Anderson & Anderson, from the same locality, showing the smaller, more slender form of the leaf (Bernard Price Institute for Palaeontology, Johannesburg, Specimen BP/2/8024). Photos by H. Anderson.

Box 8.3 | Identifying glossopterid species

Glossopterid leaves are often found complete and so, unlike with most other pteridosperms, we can see the shape of the entire leaf rather than just its building-blocks (pinnae and pinnules). Although this may seem to be an advantage, it can in fact cause difficulties as the outline of a leaf tends to be affected more by environmental factors and where it was borne on the plant, than its systematic position.

The leaves are still often classified using a system of morphogenera. *Glossopteris*, after which the group is informally named, is characterised by a strong midvein running along its length and anastomosed lateral veins. These leaves with anatomosed veins may look superficially like some angiosperm leaves, and there have been suggestions that the glossopterids may have been ancestral to the flowering plants. On detailed examination, however, the glossopterid venation is in fact quite different from that of most angiosperms, as it consists of only one order of veins; the venation of most angiosperm leaves consists of vein-meshes within vein meshes. Other, less abundant leaf-types also probably belong to the glossopterids, such as *Gangamopteris*, which does not have a midvein, and *Rhabdotaenia*, which has non-anastomosed veins.

Such morphogenera of leaves are of course largely artificial. In a few cases, however, the remains of the reproductive organs have been found attached to the leaves. In such cases, the species based on leaf morphology have been transferred to a more natural fossil genus based on the fertile organs. For instance, the glossopteroid leaf shown in Fig. 8.16 is now included within the fossil genus *Lanceolatus* because examples have been found with the characteristic ovule-bearing structures, even though the leaf has all of the morphological characteristics of *Glossopteris*. Eventually, all species of glossopterids will hopefully be assignable to one or other of these more natural fossil genera. Until we have worked out the correlation between the leaves and their reproductive structures, however, the artificial morphogenera such as *Glossopteris* will still have a role.

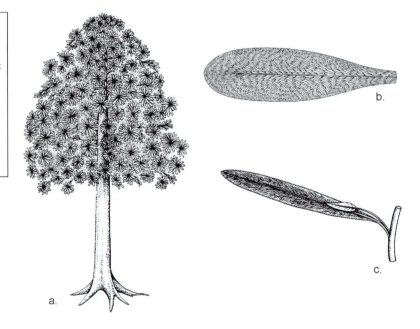

Fig. 8.16 Glossopterid plants. a. Reconstruction of a Permian-age tree bearing glossopterid leaves. b. *Glossopteris browniana* Brongniart leaf showing characteristic mesh veining, from Zeiller (1900). c. A fertile leaf of the genus *Lanceolatus* with a stalked ovule-bearing structure known as a capitulum. Drawings by A. Townsend.

smaller, depending on the genus. The head of the sporophyll-like structure may have been flat or partly enrolled around the ovules, and in one case the laminate head entirely enveloped a single ovule resembling a cupule such as is seen in some other Palaeozoic gymnosperms. The male reproductive organs are less well known, but seem to have been essentially similar to the female organs except that pollen sacs rather than ovules were attached to the sporophyll heads. Male and female organs seem to have been borne separately on different leaves, but there is no evidence that there were separate male and female plants (i.e. they were dioecious, as in modern *Ginkgo*). At one time it was thought that some of the fructifications were bisexual, which caused some palaeobotanists to speculate about the glossopterids being ancestral to the angiosperms, but this is now discounted as having been due to a misinterpretation of complex female structures.

The glossopterids' nearest relatives were probably the caytonias, a group of Mesozoic gymnosperms, which are dealy with in Chapter 9. However, the ancestral history of the glossopterids is a mystery. There were only a few gymnosperms growing at these middle palaeolatitudes during the Carboniferous Period; most of these areas were covered by glacial ice for much of the time, or supported only very restricted tundra-like vegetation. The Carboniferous palaeotropical vegetation had several groups of gymnosperm, but none had reproductive organs or foliage that are comparable. Recent discoveries in Antarctica and Australia of petrified glossopterid fossils, including reproductive organs, may help us understand better the detailed anatomy of these plants and this may in turn help resolve some of the problems surrounding their ancestry.

Cordaites

For the last group of gymnosperms to be dealt with in this chapter, we return to the Pennsylvanian and Permian palaeotropical forests. They were trees or scrambling shrubs (Fig. 8.17a–b) that occupied a range of different habitats,

including conditions similar to modern-day mangroves and drier areas in more marginal parts of the wetlands. Unlike most of the contemporary gymnosperms, at least those growing in the palaeotropics, cordaites did not have fern-like fronds, but instead had elongate, strap- or tongue-shaped leaves, with straight, parallel veins running along their length (Fig. 8.18a). These leaves were helically and densely arranged on the stems, giving a very superficial resemblance to the modern *Dracaena* tree.

The smaller branches had a zone of conifer-like secondary wood surrounding a large central zone of pith. As these branches grew and elongated, no new pith was added and as a consequence gaps appeared in the central cavity, separated by thin septae of pith. As with the horsetails (see Chapter 6) this cavity often became filled with sediment after the death of the plant, resulting in a pith cast (known as *Artisia* – Fig. 8.18c) with numerous transverse lines marking the septae of pith.

The reproductive structures consisted of complex inflorescences (Fig. 8.18b) that grew among the leaves on the most distal branches. Each inflorescence consisted of four ranks of bracts attached along a slender axis, with a small cone attached in the axil of each bract (Fig. 8.17c). The cones themselves comprised helically arranged scales, most of which were sterile, but the most distal ones bore either an ovule or a number of pollen-sacs. The ovules were flattened, with two more or less prominent lateral wings. The pollen was very like that of the early conifers, in which an air-bladder (or saccus) surrounds the pollen grain to aid wind transportation. The inflorescences were either all male or all female, but it is not known if the cordaites were dioecious (i.e. there were separate male and female plants).

From the similarities in their stem anatomy and reproductive structures, there is no doubt that these cordaites were closely related to the early conifers, and they are usually included in the same class, the Pinopsida. It is tempting to take a progymnosperm ancestor such as *Archaeopteris* (see Chapter 4) and hypothetically to elongate its spirally attached leaves, to produce a cordaites-like plant. By

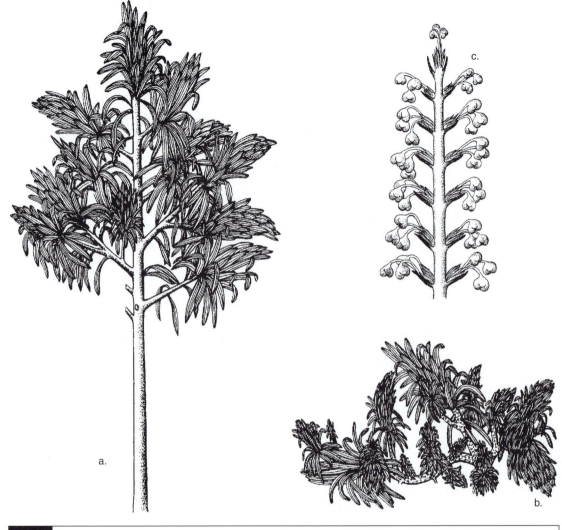

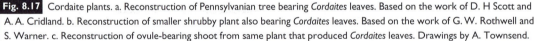

Fig. 8.17 Cordaite plants. a. Reconstruction of Pennsylvanian tree bearing *Cordaites* leaves. Based on the work of D. H Scott and A. A. Cridland. b. Reconstruction of smaller shrubby plant also bearing *Cordaites* leaves. Based on the work of G. W. Rothwell and S. Warner. c. Reconstruction of ovule-bearing shoot from same plant that produced *Cordaites* leaves. Drawings by A. Townsend.

subsequently reducing these leaves again, you could potentially get a conifer. Many palae-obotanists now believe that this is much too simple a story, but there is, nevertheless, general consensus that the cordaites and conifers were closely related groups, perhaps both derived from a common ancestor among the archaeopteridian progymnosperms.

The cordaite plants as outlined above are typical of the Pennsylvanian and Permian palaeotropical forests, but very similar foliage occurs in the contemporary temperate floras. Especially in the northern temperate palaeolatitudes, plants producing such leaves were very abundant and their remains have formed thick coal deposits in Siberia. However, the associated reproductive structures are rather different, especially in the cones not being aggregated into inflorescences. Unfortunately, few fossils of these plants have been found with anatomy preserved, at least compared with the coal ball petrifications of the palaeotropical cordaites, and so their affinities at the moment remain rather uncertain.

c.

a.

b.

Fig. 8.18 a. *Cordaites* leaf with many parallel veins (BMNH Specimen V.29275). b. *Cordaitanthus* sp., Middle Pennsylvanian ironstone nodule, Phoenix Brickworks, Crawcrook, Durham, England. In the illustration, the ovules can be clearly seen in the upper part of the specimen, on the left-hand side. BMNH Specimen V.28321. c. *Artisia transversa* Sternberg pith cast. Middle Pennsylvanian, Glamorgan, Wales. NMW Specimen 40.315G.2. Scale bars = 10 mm. From Cleal & Thomas (1999) (a, b) and Cleal & Thomas (1994) (c).

Box 8.4 | Identifying cordaite foliage

The leaves are the most commonly found macrofossils of this group of plants (Fig. 8.19). Unfortunately, however, the leaves were large and so normally all that we have are broken fragments. Equally unfortunately, these leaves were not made up of sub-units, similar to the pinnae and pinnules of the pteridosperm fronds, and so the fossils tend to be irregularly broken-up pieces of the leaf. Hence leaf-shape is rarely of use, although sometimes the shape of the leaf apex if it is preserved has been used as a character. The only other morphological character that is therefore available is the venation. This consists of the true veins made up of vascular tissue, which tend to be the thickest, and 'false veins' which are thinner and made up of strands of sclerotic tissue that presumably helped strengthen the leaf. Morphospecies have therefore tended to be recognised based on the density of the veins, and the relative arrangement of the true veins and 'false veins', sometimes combined with the shape of the leaf-apex.

Recently, the systematics of the cordaite leaves have been overturned following work on the cuticles (e.g. Šimůnek, 2007). This has shown that what appear to be natural fossil species can be recognised based on characters of the stomata and other epidermal structures, but that this bears little relationship to the species recognised on leaf shape and venation. It would seem that the latter characters are all but meaningless in any useful taxonomic sense. If cuticles are not preserved, then at present it seems that cordaite leaves will be difficult to deal with.

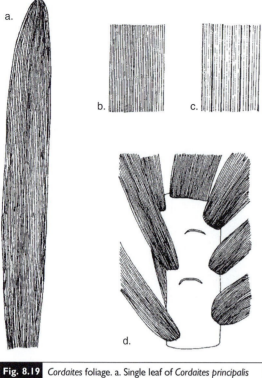

Fig. 8.19 *Cordaites* foliage. a. Single leaf of *Cordaites principalis* (Germar) Geinitz. b. Example of *Cordaites* leaf venation with all veins of equal width. c. Example of *Cordaites* leaf venation with thick veins separated by several thinner 'veins'. d. Portion of leafy shoot of *Cordaites angulosostriatus* Grand'Eury; after Grand'Eury. Drawings by D. M. Spillards. From Cleal & Thomas (1994).

Recommended reading

Anderson *et al.* (2007), Beck (1988), Chandra & Surange (1979), Cleal & Shute (1995), Cleal & Thomas (1994), DiMichele *et al.* (2006), Kerp (1988), Kerp & Haubold (1988), Long (1960), Mapes & Rothwell (1991), Pigg & Trivett (1994), Poort & Kerp (1990), Retallack & Dilcher (1988), Rothwell (1981), Rothwell *et al.* (1989), Serbert & Rothwell (1992), Šimůnek (2007), Taylor (1965), Taylor & Millay (1979, 1981), Thomas & Cleal (1993), Trivett & Rothwell (1991), Wagner (1968), White (1986).

Chapter 9

Modern gymnosperms

Many of the plant groups dealt with in Chapter 8 became extinct at the end of Permian times – apparently casualties of the great Permian/Triassic boundary (P/T) extinction event (Fig. 8.1). For some time after this cataclysmic event, terrestrial vegetation seems to have struggled and most of the fossil record from Early Triassic times consists of non-seed-plants. However, during middle and late Triassic times, we see the appearance and rapid diversification of a whole new set of seed-plants (Figs. 9.1, 11.7), such that the Triassic Period has been termed the 'Heyday of the Gymnosperms'. These groups of seed-plants remained abundant and widespread during the Jurassic Period, but many went into decline during Cretaceous times, probably through being out-competed by the newly appeared flowering plants (see Chapter 10). Some, such as the Bennettitales and Caytoniales, became extinct by the end of Cretaceous times. Others, such as the Cycadales and Ginkgoales have survived through to the present-day, but with much reduced diversity and abundance. But one group of gymnosperms has survived and flourished into modern times – the conifers that still occur as dense forests, especially in higher latitudes.

It must be emphasised that there is no natural systematic distinction between the gymnosperms dealt with in this and Chapter 8. We have chosen to divide the information between two chapters on overall stratigraphical grounds and for the convenience of the reader. Nevertheless, two of the groups discussed below have their first occurrences in the Palaeozoic

Era, while at least one of the groups described in Chapter 8 (the peltasperms) ranges through into early Mesozoic times. Nor are there any specific characters that separate the two groups of gymnosperms.

Early conifers

Conifers are the most successful gymnosperms living today, consisting of over 600 species grouped into about 70 genera and seven or eight families (examples are shown in Fig. 9.2). They have an almost worldwide distribution, but are especially common in the cooler high latitudes and altitudes where they form extensive forests. They include the largest living organisms of any type: the giant redwoods of California (*Sequoia sempervirens* (Don) Endl.). They also include some of the oldest living organisms, such as the bristlecone pines of the high mountains of south-west USA (Fig. 9.2c–d), some of which are estimated to be nearly 5000 years old. Conifers are typically substantial evergreen trees, although a few are small shrubs. Their foliage consists of helically arranged slender needle-like leaves or small fleshy scale-like leaves attached helically or in opposite and alternate pairs to the branches, and a very few have broader leaves with several veins along their length (such as the Kauri pine *Agathis* and the Monkey-puzzle *Araucaria*). The reproductive structures are clustered together into structures called cones, with separate male and female cones (Fig. 9.3). Unlike most angiosperm woods,

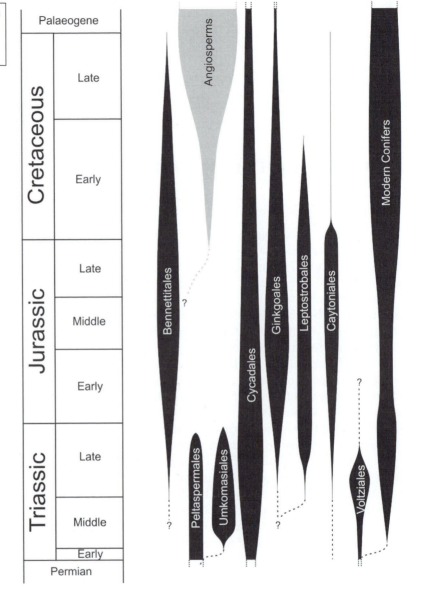

Fig. 9.1 Ranges of the main orders of gymnosperms through the Mesozoic Era. Based mainly on Anderson *et al.* (2007).

conifer woods have no vessels running along their length (Fig. 10.3).

Several types of conifer fossils are now known from Palaeozoic rocks (Fig. 8.1). While they certainly had many resemblances to modern conifers, they are usually assigned to their own order (Voltziales) due to differences in the structure of their ovulate cones. They appear to have arisen in habitats away from the lowlands where most fossils are preserved and so the fossil record of their early evolution is extremely poor.

Their pollen has been found in strata as old as Early Pennsylvanian, most notably in China. The oldest known macrofossils of the Voltziales are fragments of charcoalified conifer leaves found in Middle Pennsylvanian deposits in northern England, and are thought to have been washed down from drier habitats surrounding the depositional basin. Although extremely small, they are beautifully preserved and show what seem to be typically conifer-like stomata. Higher in the Middle Pennsylvanian Series, more complete

Fig. 9.2 Living conifers. a. *Araucaria araucana* (Molina) Koch, the monkey puzzle tree, in the Victorian Dinosaur Park in the Crystal Palace grounds, south London. This was part of the 'contemporary' planting for the Wealden part of the Secondary (Mesozoic) Island when it was constructed in the early 1850s. A reconstruction of an *Iguanodon* standing on rocks appropriate to the age of the animal and plant fossils (see Doyle & Robinson, 1993, for further information). b. *Pinus canariensis* Chr. Sm. in Tenerife. It is an endemic pine in the Canary Islands forming an open, savanna-like zone between 1200 and 2000 m with sparse shrubs and herbs. c–d. *Pinus balfouriana* Greville & Balfour, the Californian bristlecone pine, in the White Mountains in east central California at an altitude of about 14 000 feet (4500 metres). Bristlecone pines grow very slowly in isolated groves of cold temperatures, dry soils, high winds and short growing seasons. The dense, resinous wood is resistant to attack by fungi and insects. As the tree ages, much of its vascular cambium layer may die, sometimes leaving only narrow strips of living tissue connecting the roots to the remaining live branches. The average age of the trees is 1000 years with the oldest up to 5000 years old. Photos by B. A. Thomas.

Fig. 9.3 Cones of living conifers.
a. *Cedrus atlantica* (Endl.) Carr.
(cultivated) showing female cones.
b. *Pinus mugo* Turra growing in the
Tatra Mountains of Slovakia,
showing male (left) and female
(centre) cones. Photos by
B. A. Thomas.

segments of conifer-like branches are found and, in the Upper Pennsylvanian Series, the oldest anatomically preserved conifers can be found, for instance in Oklahoma, Texas and Kansas.

These early conifers probably looked rather like the modern Norfolk Island pine (*Araucaria excelsa*), with rigid branches and scale-like leaves (Fig. 9.4). Much of the interest in early conifers has focussed on the structure of their cones, especially the female cones (Fig. 9.6). Some of the earliest known cones in which anatomical detail is preserved belong to the Pennsylvanian family the Utrechtiaceae, in which the cones consisted of helically arranged bracts each with a fertile shoot in its axil. The fertile shoot consists of several helically arranged sterile scales and one or more fertile scales. Each fertile scale bears one ovule attached laterally to the scale, with its micropyle facing the axis of the dwarf shoot (i.e. it is reflexed). There are some similarities between these fertile shoots and the cones of the cordaites, but the ovules in the latter are attached terminally to the scales (see Chapter 8). The terminology of the reproductive parts is completely different between the two groups, so what was an inflorescence in the cordaites is referred to as a cone in the conifers, and the cordaite cones become fertile shoots.

Fig. 9.4 Leafy shoots of Early Permian conifers. Scale bars = 10 mm. a. *Culmitzschia laxifolia* (Florin) Clement-Westerhof, Nahe Group, Sobernheim, Germany (Laboratory of Palaeobotany and Palynology, Utrecht, Specimen 13145). b. *Culmitschia parvifolia* (Florin) Kerp and Clement-Westerhof, Lauterecken-Odernheim Formation, Oderheim, Germany (Paläontologisches Institut, Universität Mainz, Specimen Q 1569). Photos by H. Kerp. From Cleal & Thomas (1999).

Conifers and cordaites used to be thought of as directly related, but this view is not now so widely accepted because of the differences in attachment and orientation of the ovules. There are nevertheless so many points of similarity that it is difficult to imagine that they did not at least share a fairly close common ancestor.

In some other Late Palaeozoic conifers (e.g. *Emporia*) the fertile shoots were somewhat flattened (Fig. 9.6a) and in others the ovules were borne laterally rather than terminally to the fertile scale (family Majonicaceae). This trend of flattening in the fertile shoots continued with the Late Permian family the Ulmanniaceae, in which the scales became fused into a single, lobed scale with ovules attached to some of the lobes (Fig. 9.6c).

The pollen cones of these plants have not been so widely studied. However, they appear to have been simpler and more similar to each other in structure than the ovulate cones, and consisted of sporophylls arranged helically around a central axis, with two or more pollen sacs (number depending on the genus) attached to the stalk of each sporophyll.

These Palaeozoic conifers are mostly known from palaeoequatorial vegetation. Conifer-like foliage can also be found in the southern, Gondwana floras, but the associated reproductive organs are quite different from those described earlier. The ovulate cones of the Early Permian Ferugliocladaceae, best known from South America, were simple cones without sterile bracts, while in the Permian Buriadiaceae from India, the ovules were not in cones at all but directly attached to the vegetative branches, replacing normal leaves. It seems that the presence of conifer-like foliage is mainly a reflection of the conditions to which the plants had become adapted (mainly drier habitats) rather than being unequivocal evidence of a close relationship. It is likely that the modern conifer families arose from one or more of the voltzialean families of the palaeoequatorial belt but there remains considerable uncertainty as to which. Where once there seemed to be a simple story of evolution from cordaites to volzialeans to modern conifers, there now seems to be a complex phylogenetic history, only part of which is directly preserved in the fossil record.

Box 9.1 | Identifying conifer shoots

Leafy shoots are by far the most commonly found macrofossils of conifers (Fig. 9.4). Such shoots are usually very distinctive and can be immediately recognised as conifer; only lycophyte shoots can sometimes be confused with them. However, distinguishing conifer families on foliage alone is not so easy. If cuticles are preserved, this can often help, but many conifer fossils are merely impressions or poor compressions with no cuticles. To help overcome this problem, most palaeobotanists assign Mesozoic conifer foliage, unless reproductive structures are attached or there is other good evidence of their affinities, to one or other of about eight morphogenera defined exclusively on the shape, venation and attachment of the leaves (Fig. 9.5). Typical characters are whether the leaf is constricted at the base, its length relative to its width, whether the leaf sticks out from the stem, the number of veins per leaf, and the pattern of arrangement around the stem (phyllotaxy). It has been shown in some of the better-preserved fossils, that foliage of more than one morphogenus was attached to different parts of the same tree. This means that some names that appear in the literature, such as *Cupressinocladus* (the morphogenus for small, *Cupressus*-like leaves attached to the stem in decussate pairs or alternating whorls), cannot be used on their own as evidence of the family Cupressaceae. Nevertheless, provided the limitations of the scheme are kept in mind, these morphogenera provide a useful way of recording conifer foliage found in the fossil record, without making any unwarranted claims as to their affinities.

Palaeozoic conifer shoots present greater difficulties as they belong to families that are now extinct. It is usual to refer such shoots without cuticles to the morphogenus *Walchia*, whereas those that have cuticles showing epidermal anatomy are referred to *Culmitzschia*. However, even if epidermal anatomy is shown, it is difficult to assign the shoot to a particular family, the latter having been defined essentially on the structure of the ovule-bearing cones. Unless evidence of the cones is available, it is all but impossible to classify the shoots in any systematically meaningful way.

Fig. 9.5 Morphogenera of some Mesozoic conifer shoots. a. *Brachyphyllum* with leaves contracting gradually from broad bases; leaves less than five times as long as broad with a free portion shorter than the width of its basal part. b. *Lindleycladus* with leaves contracting basally and then expanded to form a blade; leaf lamina with several veins. c. *Pagiophyllum* with leaves contracting gradually from broad bases; leaves less than five times as long as broad with free portion longer than the width of its basal part and diverging from stem. d. *Geinitzia* with leaves contracting gradually from broad bases; Leaves at least five times as long as broad and round or rhomboidal in section. e. *Elatocladus* with leaves contracting gradually from broad bases; leaves less than five times as long as broad with free portion longer than the width of its basal part, thin, flat and, in some species, twisted into a horizontal plane. Drawings adapted from Kendal (1952) and van Konijnenburg-van Cittert & Morgans (1999).

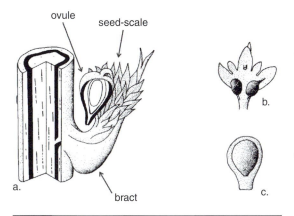

ovule
seed-scale

a.

bract

b.

c.

Fig. 9.6 a. Bract and fertile shoot from the cone of *Emporia lockhardii* Mapes & Rothwell, from the Pennsylvanian Subsystem of Kansas, USA. b. Seed scale with two ovules from the cone of *Pseudovoltzia liebiana* (Geinitz) Florin from the Late Permian of Germany. c. Seed scale with a single ovule from the cone of *Ullmannia bronnii* Goeppert, from the Upper Permian of Germany. From Thomas & Spicer (1987).

Modern conifers

As with the arborescent lycophytes and many ferns, most of the voltzialean conifer families became extinct at the P/T (Permian-Triassic) boundary. Only some of the Ulmanniaceae seem to have survived through until Triassic times, eventually becoming extinct in early Jurassic times. Instead, many new conifer families appeared during the Triassic Period, such as the Pinaceae, Araucariaceae, Cupressaceae, Podocarpaceae and Taxaceae (Figs. 9.1, 11.5). These families lived throughout the rest of the Mesozoic Era, survived the Cretaceous/Tertiary (K/T) boundary extinction event, continued through the Cenozoic Era (e.g. Fig. 9.7) and are still alive today. The Taxaceae, which includes the yews, has sometimes been regarded as belonging to a different order (Taxales) partly because the ovules are not in cones, but are

a. b. c.

Fig. 9.7 Mesozoic and Cenozoic conifers. a. Ovule-bearing cone attached to a shoot of *Elatides williamsonii* (Brongniart) Nathorst. Scale bar = 10 mm. Middle Jurassic Grisethorpe Plant Bed, Cayton Bay, Yorkshire, England (NMW Specimen 84.27G.361). Photo by NMW Photography Department. b. *Araucarites* sp. shoot, Eocene London Clay, Sheppey, England. Scale bar = 5 mm. Photo by M. E. Collinson. c. *Pinus dixonii* (Bowerbank) Gardner cone, Palaeogene Barton Clay, Barton, England (BMNH Specimen V.60468). Scale bar = 10 mm. Photo by NHM Photography Department. From Cleal & Thomas (1999) and Cleal et al. (2001).

Fig. 9.8 Silicified cones of *Araucaria mirabilis* (Spegazzina) Windhausen from the Jurassic Cerro Cuadrado Petrified Forest in Patagonia, Argentina. Scale bar = 10 mm for both photos. a. An immature cone with attached peduncle. b. Section showing the bracts and ovules surrounding the central axis. From Stockey (1978), with permission.

single and surrounded by a red fleshy structure called an aril. *Taxus* is also unlike many other conifers in being dioecious (there are separate male and female plants). However, modern phylogenetic analyses suggest that *Taxus* is a true conifer and so we have retained it within the same order as the rest of the conifers.

As with the Palaeozoic conifers, these more modern conifers are difficult to identify on foliage alone. However, evidence of fructifications that are clearly indicative of these families has been found in Jurassic rocks, especially in the Middle Jurassic flora of Yorkshire (e.g. Fig. 9.7), and in some cases as far back as the Triassic Period. The Araucariaceae are represented by well-preserved petrified cones from at least Jurassic times (Fig. 9.8). It appears that the general trend observed in the female cones of Palaeozoic conifers continued with the Mesozoic conifers, with a general reduction in the number of ovules attached to each ovuliferous scale, and a tendency for the ovuliferous scale to become more fused to the sterile bract. These features can often be determined in the fossils, especially if the cones can be macerated. Phylogenetic changes in the male cones are less easy to determine as they tended to be less robust than the ovuliferous cones and thus the details are often not as well preserved. Here too, however, there seems to have been a general

trend towards simplification in the arrangement of the male cones, although there is some variation in how the pollen-sacs are arranged on the bracts.

Much attention has been paid to the origins and evolution of modern conifer families, in particular trying to integrate the evidence from the fossil record with the new phylogenetic models being proposed on molecular data. The evidence seems to be in agreement that the conifers (including the Taxaceae) are a monophyletic group and that they arose, probably in an explosive radiation, during Triassic times. However, there is still little consensus about the relationships between the families. Other Mesozoic conifers show a combination of characters of the modern families: for instance, the misleadingly named *Pararaucaria* from the Middle Jurassic Cerro Cuadrado flora of Patagonia refers to petrified cones that combine features of the Cupressaceae and Pinaceae. Clearly, much needs to be done before this vexed problem of conifer phylogeny will be resolved.

Most Mesozoic conifer families have survived to the present-day, but there is one notable exception: the Cheirolepidiaceae (Fig. 9.9). This appears to have been one of the most abundant of the Mesozoic conifers, with an almost worldwide distribution from Late Triassic to middle Cretaceous times. The pollen (known as *Classopollis*), in particular, is highly distinctive, with a trilete mark, and a groove that extends around the grain about half way between its equator and the end furthest from the trilete mark. The cones had large, lobed ovuliferous scales, which were separate from the large bracts. The scales were shed leaving cones consisting only of the helically arranged bracts, which are often found fossilised. Why the cheirolepidiaceans became extinct is still a mystery. It certainly had nothing to do with the K/T extinction event, as the family seems to have already disappeared by middle Cretaceous times. Perhaps it reflects increased competition from the more modern conifer families and the angiosperms, which seem to have been diversifying at about that time. There may also be an element of habitat loss, as middle Cretaceous times saw rising sea levels, which may have flooded the flat coastal fringes where many cheirolepidiaceans were growing (Fig. 9.10).

Fig. 9.9 Reconstruction of the type of cheirolepidiacean conifers that were important components of the Mesozoic forests. The inset shows an ovuliferous cone attached to a leafy shoot; the left cone has its seed-bearing scales shown still attached, the right cone has shed the scales and its seeds. Paintings by P. Dean, based on the work of J. E. Francis.

Ginkgoales

Ginkgo biloba L. is probably the best example of a 'living fossil' plant (Fig. 9.11). In the Mesozoic Era, the ginkgophytes were a highly diverse group with an almost global distribution, but today there is just this single species. It was once thought to be totally extinct in the wild, the last remaining examples having been pre-served in Japanese and Chinese temple gardens, where it was discovered by western botanists in the 1700s. However, a small stand of wild *Ginkgo* has since been found in Zhejiang Province, south-east China, and further stands may also be growing in some of the less well known parts of that large country. Why *Ginkgo* under-went such a dramatic decline is far from certain. Because it is such an elegant tree with attractive foliage, it has been widely cultivated in Europe and North America, where it has proved to be a remarkably resilient and adaptable plant; hardly what one would expect of a species on the verge of extinction. It nevertheless seems that the ginkgophytes suffered as a result of competition with angiosperms, and became more and more restricted to northern temperate forests during Cenozoic times. It may have had something to do with the plants being dioecious and relying on wind for pollination. When ice spread over many parts of these temperate forests during the last Pleistocene ice age, the main remaining habitats of the *Ginkgo* were destroyed.

Ginkgo-like foliage (e.g. Fig. 9.12) is known from the Upper Triassic Series. There have even been putative Palaeozoic ginkgophyte fossils, although most of these are now disputed. The problem with elucidating the evolutionary his-tory of ginkgophytes is that relatively few repro-ductive structures are known. Some, such as the Middle Jurassic *Allicospermum* from Yorkshire, was very similar to living *Ginkgo*: ovules were borne at the ends of stalks, usually in pairs, and attached to short-shoot scales or leaves. Other, mainly Jurassic ginkgophyte ovule-bearing struc-tures were very different from those of modern ginkgos and their parent plants have now been assigned to separate families. The Karkeniaceae, for instance, had numerous ovules borne in

Fig. 9.10 'Birds' nests' on the Isle of Portland in southern England are Mesozoic-age stromatolites that were produced by the activity of cyanobacteria around the stumps of cheirolepidiacean tree trunks that decayed soon afterwards. The trees were rooted in the Great Dirt Bed palaeosol below. The stromatolites were growing mats of bacterial filaments whose photosynthetic activity depleted carbon dioxide in the surrounding water. This depletion initiated the precipitation of layers of calcium carbonate over the bacterial filaments. Photo by B. A. Thomas.

densely packed strobili (cones), whilst the Schmeissneriaceae had the ovules attached laterally to a main axis. All but the Ginkgoaceae became extinct during Cretaceous times.

Students of living plants have tended to regard the Ginkgoales as closely related to conifers, a view apparently being supported by at least some molecular (DNA) evidence. However, some palaeobotanists have argued that the Ginkgoales are related to the Palaeozoic cordaites or even the peltasperms. The difficulty is, of course, that these extinct groups cannot be incorporated into the molecular-based phylogenies, although morphology-based phylogenetic analysis might provide insights into the problem.

Cycads

With eleven genera and some 185 species, this is one of the most diverse groups of living gymnosperms (Fig. 9.13), second only to the conifers. Cycads have a wide but rather unusual distribution, with each genus being restricted to a particular geographical area; no genus (with the possible exception of *Cycas*) occurs in more than one continent. It is now thought that these isolated genera are the remnants of much more widely distributed Mesozoic and early Tertiary populations, and indeed the fossil record indicates that the cycads were much more abundant and diverse, especially in Mesozoic times.

Modern cycads usually have an unbranched main stem, covered with a thick layer of persistent leaf-bases left after the shedding of old leaves. The stem may be long and thick, giving the plant a superficial resemblance to a palm-tree (e.g. *Macrozamia moorei* Muell), short and squat (e.g. *Cycas revoluta* Thunb.) or may be short and exclusively underground, like a tuber (e.g. some *Zamia* species). The leaves are large, compound fronds, which are usually rather thick and stiff. Some cycads are extremely long lived, being hundreds of years old.

One of the distinctive features of cycads is the way that the vascular tissue enters the leaf from the stem. A strand of vascular tissue (a leaf trace) branches off from the stele of the stem but does not immediately enter the petiole of a leaf. It instead divides into two strands, which pass around either side of the stem. These strands eventually enter the base of a leaf on the opposite side of the stem from where the leaf trace left the stele. This distinctive feature known as 'girdling leaf traces' has been recognised in fossils as old as Triassic and seems to be unique to the cycads.

The Cycadales are usually divided into two or three families. All are dioecious and there is no evidence to suggest that any ancestral forms were not. Most cycads have compact female cones, with a central axis bearing woody peltate megasporophylls, each with a pair of ovules. The stem ceases growth when forming a cone and is only re-initiated by a new meristem forming

Fig. 9.12 A ginkgophyte leaf *Ginkgoites pluripartita* (Schimper) Seward. Scale bar = 20 mm. Lower Cretaceous (Wealden) of Osterwald, Germany (Geologisch-Paläontologisches Institut und Museum der Georg-August-Universität, Göttingen, Germany, Specimen P4–5). Photo by J. Watson. From Cleal & Thomas (1999).

Fig. 9.13 Living cycads. a. *Cycas* plant, Cairns Botanic Garden, Australia. Photo S. G. Harrison. b. *Zamia*, in San Salvador, with ripe female cones that are disintegrating and exposing the seeds for animal dispersal. Photo by B. A. Thomas. c. *Cycas* with a spiral of megasporophylls in the centre of the foliage leaves. Seeds are visible on most of the megasporophylls to the right. Photo by B. A. Thomas.

near the cone's pedicel. This pushes the cone to the side giving an appearance of uninterrupted growth. Traditionally, these cycads have been assigned to the family Zamiaceae, although some authors separate off the extant *Stangeria* and *Bowenia* into a separate family, the Stangeriaceae, based mainly on differences in the foliage. *Cycas* species are rather different from the others in producing sporophylls in place of the leaves. The upper part of the sporophylls are leaf-like and the ovules are attached to the rachis-like basal area, and are placed in their own family, the Cycadaceae. All the cycads, including *Cycas*, have compact male cones.

Cycads have many similarities with the Medullosales described in the previous chapter, and it is likely that they share a common ancestry. Evidence of what might be cycad pollen has been found in Pennsylvanian deposits, probably having blown into the lowland depositional basins from plants growing in more 'upland' (extra-basinal) habitats. The oldest known cycad macrofossils are of Early Permian age, from China (Fig. 9.16a). These include fronds and, most significantly, large terminally divided sporophylls that are very similar to those seen in the cones of living *Cycas*. Other lines of evidence have suggested that *Cycas* represents the most primitive of the living genera of cycads, and the Chinese evidence seems to support this view. There are some other Palaeozoic fossils (*Lesleya*, *Phasmatocycas*) that have been interpreted as primitive cycads but this interpretation has now been questioned and they have been assigned to a separate order, the Phasmatocycadales – a group probably related to the Cycadales, but not ancestral to it.

The problem with the fossil record of the cycads is that many species seem to have favoured drier habitats, far away from where they were likely to be fossilised. Nevertheless, the fossil record strongly suggests that cycads reached their zenith during the Mesozoic Era, especially between Late Triassic and Early Cretaceous times. Not all cycad-like frond fragments found in rocks of this age are true cycads (some belong to the Bennettitales, dealt with in the next section). There are nevertheless abundant cycad remains in many floras of this age, including leaves (Fig. 9.16) with their distinctive cuticles and numerous fructifications. Despite this abundance, we are still a long way from confidently being able to reconstruct whole plants for these Mesozoic cycads. Published reconstructions have tended to make them look like modern cycads, with a stout, scaly trunk bearing an apical crown of fronds. There have also been reconstructions showing them with a long, slender trunk, somewhat resembling tree ferns (Fig. 9.17a). The suggested reconstruction of what has become known as the *Beania* tree (Fig. 9.17b) shows the fronds being borne on a branching stem, although the author of this reconstruction (the eminent palaeobotanist Professor Tom Harris) confessed that the evidence for such a branched stem was not strong.

The relationship of these Mesozoic cycads to the living forms is not always clear. Some cones had fimbriate, *Cycas*-like sporophylls, similar to those found in Permian macrofloras. However, there are also ovulate cones with peltate sporophylls attached to a central axis more reminiscent

Box 9.3 | Identifying cycadophyte foliage

Both cycad and bennettite remains occur abundantly in Mesozoic sediments, and they share so many similarities that they can easily be confused. The bennettites had very similar shaped leaves to the cycads and these are very difficult to distinguish on the basis of morphology alone. However, if cuticles are available, the two types of leaves can be easily told apart. The stomata on the lower cuticle of cycad leaves consist only of a pair of guard cells surrounding the pore, whereas bennettites have a second pair of specialised cells (known as subsidiary cells) surrounding the guard cells (Figs. 9.14). The upper cuticle of bennettite leaves is also quite distinctive, being rather brittle and showing epidermal cells with very sinuous walls. These differences have proved consistent between these two groups of plants, allowing even quite small fragments of leaf to be distinguished.

Unlike many of the pteridosperm fronds discussed in Chapter 8, cycadophyte fronds tend to be relatively small in size and it is not unusual to find fossils of entire leaves or large segments of leaves. This makes their classification somewhat easier, especially if cuticles are also available. Various fossil genera of cycadophyte foliage have been defined, based on characters such as whether the leaf is entire or pinnately divided, the size and shape of the ultimate segments (pinnae) and the veining (Fig. 9.15). Most of the fossil genera defined on such characters appear to be exclusively bennettitalean (e.g. *Otozamites*) whereas some are cycadalean (e.g. *Pseudoctenis*). There is a problem with the entire, undivided cycadophyte leaves, examples of which occur in both the bennettitaleans (*Nilssoniopteris*) and cycadaleans (*Nilssonia*). Unless cuticles are available it is impossible to assign a particular fossil to one or other of these genera, and so a separate morphogenus is used for entire cycadophyte leaves that lack cuticles: *Taeniopteris*. This appears to be an exception, however, and morphology can, on the whole, be used as a reasonable guide to the identification of such leaves.

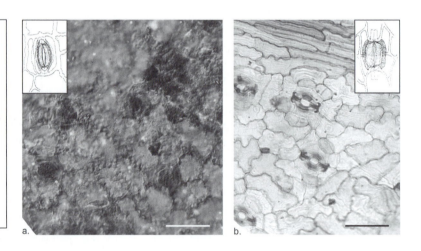

Fig. 9.14 Cuticles of cycadophyte leaves. Both scale bars = 50 μm. Inserts show drawings of stomata, from Cleal *et al.* (2001).
a. *Nissoniopteris vittata* (Brongniart) Harris, (Bennettitales), Middle Jurassic flora of Yorkshire, England. Cuticle photographed using Normarski Contrast. (NMW Specimen 90.5G.3). Photo by C. J. Cleal. b. Cleared leaf of the living *Stangeria eriopus* (Kunze) Baillon, endemic to South Africa. Photo by B. A. Thomas.

of the living Zamiaceae and Stangeriaceae, although usually rather less compact and woody. Moreover, if they can be proved to have branching stems, this would be quite different from anything known in modern-day cycadaleans. For this reason, some of these Mesozoic cycads have been placed in a separate family, the Nilssoniaceae, but there is still no consensus on this.

Fossil cycads are known through much of the Palaeogene and Neogene Systems, but were clearly undergoing a significant decline from their Mesozoic heyday. As far as one can make out, this had little to do with the K/T event and was probably more a function of the cycads (like the ginkgos) being out-competed by the angiosperms and to a lesser extent by the conifers. Most cycads are extremely slow-growing plants, taking a long time to reach maturity. This can have its advantages in certain types of habitat, but has significant disadvantages when faced with competition from fast-growing and early maturing plants such as the angiosperms.

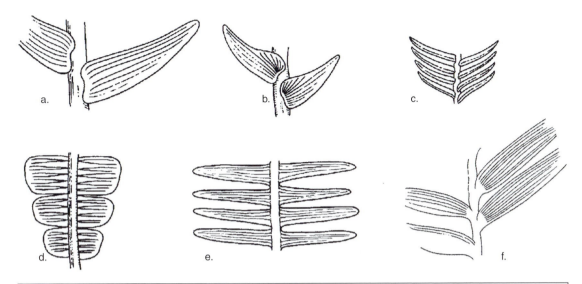

Drawings of pinnae from some of the cycadophyte fossil genera found in the Middle Jurassic floras of Yorkshire, England. a. *Zamites*. b. *Otozamites*. c. *Ptilophyllum*. d. *Anomozamites*. e. *Pterophyllum*. f. *Pseudoctenis*. a–e are bennettitalean fronds, and f, a cycadalean frond. Drawings adapted from van Konijnenburg-van Cittert & Morgans (1999).

Cycad fossils. a. *Crossozamia chinensis* Gao and Thomas, the ovule-bearing sporophyll of an early cycad cone. Scale bar = 10 mm. Lower Permian of Simugedong, Taiyuan East Hill, Shanxi, China (NMW Specimen 98.24.G7). b. Two species of *Nilssonia* leaves. Scale bar = 10 mm. The undivided or only shallowly divided leaves with a fine venation belong to *N. tenuinervis* Seward (running vertically); those with more deeply divided leaves and coarser venation belong to *N. compta* (Phillips) Bronn (running diagonally). Both species have yielded cuticles showing that they were fronds of cycads. Middle Jurassic Gristhorpe Plant Bed at Cayton Bay, Yorkshire, England (NMW Specimen 76.7G.215). From Cleal & Thomas (1999).

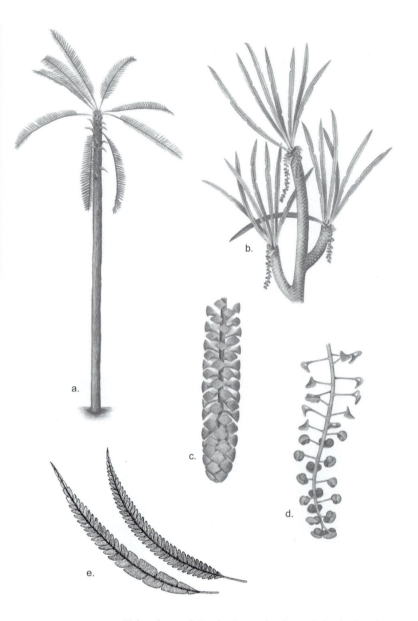

Fig. 9.17 a. The Triassic cycad, *Leptocycas gracilis* Delevoryas and Hope has been estimated to be *c.* 1.5 m tall. Based on the work of T. Delevoryas and R. C. Hope. b. Branch of the *Beania* tree from the Middle Jurassic flora of Yorkshire, England. Based on the work of T. M. Harris (1961). c. *Androstrobus* pollen-producing cone of the *Beania* tree. d. *Beania*, the ovule-bearing cone of the *Beania* tree. e. Leaves of the Middle Jurassic cycad *Nilssonia*, based on drawings by T. M. Harris. Each leaf was *c.* 0.5 m long. Paintings (a–d) and drawing (e) by A. Townsend.

Bennettitales

As stated previously, some (arguably the majority) of 'cycad' fronds found in Mesozoic macrofloras belong to a quite different group of plants, the Bennettitales (e.g. Fig. 9.18). Some authors still refer to the Jurassic Period as the 'Age of Cycads', whereas it is probably better referred to as the 'Age of Bennettites'. Some bennettitalean plants had similar stems to cycads, being covered by a thick layer of persistent leaf-bases (Fig. 9.19a) although others (especially the Triassic and Early Jurassic forms) had slender, branching stems without the layer of leaf-bases. Anatomically, there are again superficial similarities with cycad stems, with a thick central pith and a ring of secondary wood. Significantly, however, bennettite stems did not have the girdling leaf traces that are such a characteristic feature of the cycads.

These two groups of cycadophyte fronds were initially distinguished because of differences in their epidermal anatomies (see Box 9.3). However, it was subsequently realised that

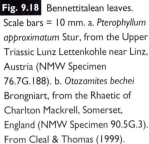

Fig. 9.18 Bennettitalean leaves. Scale bars = 10 mm. a. *Pterophyllum approximatum* Stur, from the Upper Triassic Lunz Lettenkohle near Linz, Austria (NMW Specimen 76.7G.188). b. *Otozamites bechei* Brongniart, from the Rhaetic of Charlton Mackrell, Somerset, England (NMW Specimen 90.5G.3). From Cleal & Thomas (1999).

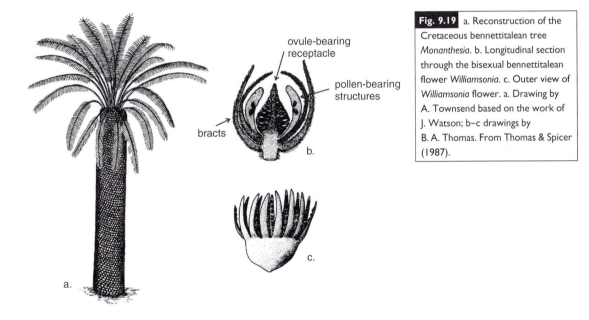

Fig. 9.19 a. Reconstruction of the Cretaceous bennettitalean tree *Monanthesia*. b. Longitudinal section through the bisexual bennettitalean flower *Williamsonia*. c. Outer view of *Williamsonia* flower. a. Drawing by A. Townsend based on the work of J. Watson; b–c drawings by B. A. Thomas. From Thomas & Spicer (1987).

ovule-bearing receptacle

pollen-bearing structures

bracts

they were the fronds of two groups of plants that were only distantly related, with fundamentally different reproductive structures. Some of the early bennettitaleans had separate ovule- and pollen-bearing organs. Most, however, had bisexual, flower-like reproductive structures (Fig. 9.19b–c). These consisted of a central, dome-shaped body of tissue (similar to the receptacle of angiosperm flowers) on which were attached numerous small ovules separated by scales. The ovuliferous part was surrounded by a sheath of pollen-bearing sporophylls, and these in turn were surrounded by a sheath of protective bracts. In the earlier bisexual bennettitalean flowers, the pollen-bearing sporophylls were able to open out to facilitate wind dispersal. In the later species, however, the sporophylls were unable to open out. This undoubtedly provided protection to the pollen-producing bodies from predation but also constrained the plants to self-pollination. This latter fact has been suggested as a major reason for the extinction of the bennettitaleans in late Cretaceous times.

The bennettitaleans had an almost world-wide distribution in the Mesozoic Era. The earliest unequivocal examples are of Late Triassic age, from Austria. They were extremely abundant throughout Jurassic and Early Cretaceous times, but declined in abundance and diversity in the Late Cretaceous, to disappear at the K/T boundary. This pattern of diversity suggests that the bennettitaleans suffered at the expense of the angiosperm diversification towards the end of the Mesozoic.

The bisexual flowers in some bennettitaleans have caused some palaeobotanists to look at the group as possible ancestors of the Gnetales (see below) and angiosperms. The exact relationship between these groups remains contentious and different phylogenetic analyses are still producing different results. However, most (but not all) authors agree that the bennettitaleans, Gnetales and angiosperms probably arose from the same plexus of plants probably sometime in very early Mesozoic, or possibly even Late Palaeozoic times.

Caytoniales

Although the bennettitaleans are the most widely distributed of the extinct Mesozoic gymnosperm groups, others have been found such as the *Caytonia*-bearing plant (Fig. 9.20). Formally, the name *Caytonia* refers to female reproductive organs occurring widely in the Mesozoic northern hemisphere, and which consisted of two rows of ovule-bearing stalked cupules attached to either side of an axis, forming an essentially planated structure. Each cupule had a more or less round outline, with an opening near its attachment to the stalk, and containing between 8 and 30 ovules, depending on the species. It was originally thought that a lip at the opening of the

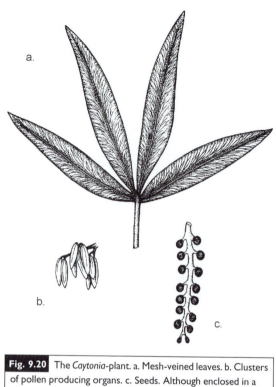

Fig. 9.20 The *Caytonia*-plant. a. Mesh-veined leaves. b. Clusters of pollen producing organs. c. Seeds. Although enclosed in a carpel-like capsule and such fossils have been widely found in Mesozoic rocks, we still do not know the overall form of the plant that produced them; whether it was a tree, a shrub, a vine or a herb. Drawings by A. Townsend, based on the work of T. M. Harris.

cupule acted like the stigma in angiosperm flowers, with pollen landing on the lip and entering the cupule via pollen tubes. However, it is now thought that pollination took place via a pollination drop, as with most gymnosperms.

Associated with *Caytonia* are clusters of elongate pollen-bearing structures (known as *Caytonanthus*) producing the same type of pollen found in the micropyles of the ovules. There are also palmate clusters of three to six (usually four) leaflets at the end of a stalk. The leaflets, which have a prominent midvein and reticulate lateral veins, are very distinctive and can usually be identified from even relatively small fragments. Using these leaf fossils, called *Sagenopteris*, it has been possible to demonstrate the wide distribution of this group of plants from Late Triassic to Late Cretaceous times in North America, Europe, and central and eastern Asia.

The *Caytonia*-bearing plants attracted much attention in the earlier part of the twentieth century because they were thought of as possible ancestors of the angiosperms. This was on the basis of the resemblance of their cupules to angiosperm ovaries, and the reticulate veining of the leaves. Both similarities are now recognised to be superficial, and the caytonias probably had little to do with the flowering plants. There is now some evidence that they may have been descendants of the glossopterids, which dominated much of the Permian vegetation in the southern hemisphere.

Other gymnosperm groups

In addition to groups discussed in this and in Chapter 8, there is a range of less well-understood gymnosperms. For instance, the Umkomasiales (sometimes alternatively known as the corystosperms) had one or two ovules borne in cupules that were superficially similar to *Caytonia* but the cupules were in a looser, non-planated configuration. The foliage of these plants, known as *Dicroidium* (Fig. 9.21), occurs extensively in Mesozoic rocks of the Gondwana continents. Other enigmatic Gondwana gymnosperm groups include the Hamshawviales, Petriellales and Matatiellales. Many of these plants have only been partially reconstructed and anatomical details are often poor or unknown, which makes it very difficult to classify them. Some authors have combined them with the ginkgos into a loose class of gymnosperms, based essentially on the fact that the ovules were borne in cupules. However, this becomes difficult to support if ginkgos are linked with conifers. Clearly, much more work needs to be done on these enigmatic seed-plants.

Gnetales

This order is based on three genera of highly specialised living gymnosperms, *Gnetum*, *Ephedra* and *Welwitschia* (Fig. 9.22). Since they show striking differences in habit and geographic distribution it may be better to separate them into families of their own. The group has attracted considerable interest from botanists as the plants appear to show features intermediate between gymnosperms and angiosperms. All of them have wood with vessels, but these are structurally different from angiosperm vessels. The end plates in gnetalean vessels are foraminate or simple, suggesting that they evolved from pitted tracheids, while angiosperm vessels are generally accepted to have evolved from tracheids with scalariform thickening. Some species of *Gnetum* also have sieve tubes that are similar in appearance to those of angiosperms, but they develop from two cambial cells rather than from one, as happens in angiosperms. *Gnetum*

Fig. 9.21 *Dicroidium dubium* (Feistmantel) Gothan, Upper Triassic Molteno Formation, South Africa (National Botanical Institute, Pretoria, Specimen PRE/F/170). Foliage of Umkomasiales plants that were one of the dominant components of Late Triassic vegetation of the southern palaeolatitudes. Scale bar = 10 mm. Photo by H. Anderson.

Fig. 9.22 Living Gnetales. a–b. *Welwitschia mirabili* Hook. f. in Namibia. a. Female plant. b. Male plant. c. *Gnetum* sp. foliage. d. *Gnetum* sp. female cone. Photos by L. M. Warren (a, b) and B. A. Thomas (c, d).

also has oppositely arranged leaves similar to those of dicotyledonous angiosperms in having broad laminae and reticulate venation. All are dioecious and have their ovules in cones that are somewhat flower-like. This, together with the angiosperm-like conducting cells suggested that they might have been ancestors of the flowering plants. Nevertheless, all three are now firmly regarded as gymnosperms. The other factor against them being angiosperm ancestors is that fossil remains that could be assigned to the

Gnetales are extremely rare. The only fossils referable to the Gnetales are some pollen grains that resemble those of *Ephedra* and *Welwitschia* from rocks from the Triassic onwards and a few debatable Tertiary stem fragments resembling *Ephedra*. There are also a number of enigmatic seed-plants, mainly of Triassic age, which may be related to the Gnetales (e.g. *Dinophyton*, *Fraxinopsis*, *Dechellyia* and *Eoantha*) but none are well enough known to be certain as to their affinities.

Recommended reading

Anderson & Anderson (2003), Anderson *et al.* (2007), Beck (1988), Florin (1951, 1963), Francis (1984), Gao & Thomas (1989), Harris (1961b–1979, 1976, 1979), Harris *et al.* (1974), Hori *et al.* (1997), Kerp (1996), van Konijnenburg-van Cittert & Morgans (1999), Mapes & Rothwell (1991), Meyen (1997), Miller (1977, 1982), Retallack & Dilcher (1988), Stockey (1988), Thomas & Cleal (1998), Watson & Sincock (1992).

Chapter 10

Angiosperms

Flowering plants, or angiosperms as they are more properly called, have dominated most terrestrial ecosystems of the world for the last 100 million years. There are nearly a quarter of a million living species in a bewildering variety of forms (Fig. 10.1) that have been grouped into between 300 and 400 families. Angiosperms range from very small floating plants such as *Lemna* to enormous, long-lived trees: they show a greater range of growth habits and morphological variation, and live in a greater range of environments than any other group of living vascular plants. Flowering plants form the basic diet of most herbivorous animals and we rely on them for agriculture, horticulture, many pharmaceutical products, fuel and building materials. Their importance to us has inevitably caused much interest in the origins and early evolution of angiosperms, and there has been considerable work done on their fossil history.

What makes an angiosperm?

It is difficult to define what actually an angiosperm is, especially if we are looking at plant fossils. The commonly held concept of a plant with showy and obvious flowers is not tenable for many angiosperms, and the range of characters recognised in living angiosperms is huge in such an enormous group with so much plasticity of form. Nevertheless, there are some features, which together can be taken to prove angiosperm affinity.

Presence of a flower.
Enclosed ovules or seeds.

A double protective layer made up of integuments surrounding the embryo sac.
Wood with vessels.
Development of the food conducting phloem.
Multi-layered pollen walls consisting of pillar-like structures called collumellae, supporting an outer covering called the tectum.
Reticulate venation pattern in the leaves.
The distinctive double fertilisation mechanism where two male nuclei fuse with nuclei in the female egg cell. A pollen tube grows through the micropyle, enters the ovule and releases two sperm cells. One fuses with the egg cell to form the diploid embryo, while the other fuses with the binucleate cell formed from the two polar nuclei to form a triploid cell that will give rise to the triploid endosperm (the food store). The presence of the double fertilisation mechanism is often given as the main argument for their monophyletic origin, although it is also argued that such an evolutionary event could quite possibly have occurred more than once in the last 100 million years.

In practice, the most commonly used diagnostic features of angiosperm fossils are vessels in wood, the reticulate venation of leaves, and pollen, even though none is exclusively found in angiosperms and not all angiosperms have them. The extant gymnosperms *Gnetum*, *Ephedra* and *Welwitschia* have vessels, while some extant angiosperms such as *Tetracentron* (Hamamelideae) and some of the Winteraceae (Magnoliales) have

Fig. 10.1 Living angiosperms. a. *Liriodendron tulipifera* L. (Tulip Tree – cultivated; Magnoliaceae). b. *Hamamelis mollis* Oliv. (witch hazel; Hamamelidaceae). c. *Magnolia* sp. flowers showing the central, elongated gynoecium surrounded by stamens and petaloid perianth (Magnoliaceae). d. *Salix alba* L. (willow; Salicaceae). e. *Laburnum* sp. (Fabaceae). e. *Alnus glutinosa* L. (black alder; Betulaceae). f. *Castanea* (chestnut; Fagaceae). Photographed in Wales by B. A. Thomas (c, f) and P. Russell (a, b, d and e).

none. *Gnetum* also has laminated leaves with reticulate venation, as do some extant ferns (e.g. *Hausmania*) and Mesozoic gymnosperms (e.g. *Sagenopteris*) while many angiosperms do not. Some angiosperms have non-tectate pollen, while some conifers, notably the Mesozoic Cheirolepidiaceae, are known to have had tectate pollen.

Flowers and some seeds are the most distinctive organs, although much rarer as fossils. Characteristics of the flowers (Fig. 10.2) form the basis of the classification system of extant angiosperms and so flowers are especially important for determining the family affinities of extinct angiosperms. Flowers are likely to be incompletely preserved, however, because petals, sepals and bracts may be shed from the living plant, lost during transportation of the flower into sediments, or simply not preserved during fossilisation. The earlier the age of any fossil flower, the more likely it is to show characteristics that fall outside the accepted range in a genus, or even family, of living angiosperms.

Angiosperms are divided into dicotyledons and monocotyledons, on the basis of their seeds having two or one embryonic cotyledons, or seed leaves. Other differences between the two groups are summarised in Table 10.1.

Table 10.1	Characters that distinguish monocotyledon from dicotyledon angiosperms.	
	Monocotyledons	Dicotyledons
Primary vascular system	Scattered	Forming a ring
Vascular cambium	Absent	Present
Leaf venation	Parallel or striate	Reticulate or net
Number of floral parts	Usually three	Four, five or many

Wood

This is the secondary xylem of conifers and angiosperm dicotyledons. Monocotyledons do not form wood when they secondarily thicken, relying instead on tough fibrous tissues to support their stems. Dicotyledon wood is often referred to as 'hard wood' and that of conifers as 'soft wood' reflecting their general hardness (Table 10.2 summarises the differences between

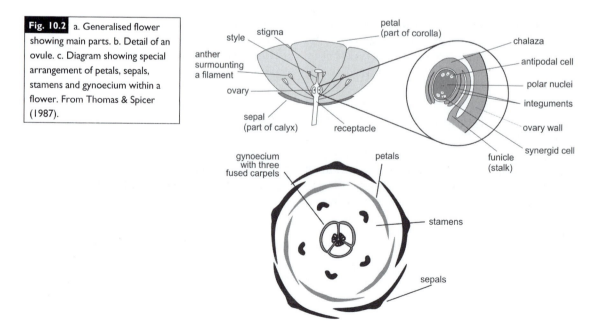

Fig. 10.2 a. Generalised flower showing main parts. b. Detail of an ovule. c. Diagram showing special arrangement of petals, sepals, stamens and gynoecium within a flower. From Thomas & Spicer (1987).

Box 10.1 | Identifying fossil wood

Fossil wood can be preserved in a number of ways (Fig. 10.3). Sometimes wood can be found in an almost natural and unmineralised condition in late Tertiary and Holocene sediments and peat. Such specimens need to be protected from drying out too fast and from fungal attack. There are fusinised (charcoalified) woods, but they are generally more fragmentary. Wood can be petrified either by silicification or calcification. Others may be pyritised. The methodology for studying fossilised woods therefore depends entirely on the method of preservation.

Any investigation of fossil wood is best started by examining a thin transverse section to decide if there are any vessels. If so, it is most probably angiosperm wood although members of the Gnetales also have vessels. Conversely, it is not automatically non-angiosperm wood if there are no obvious vessels because members of the Winteraceae, Trochodendraceae and Tetracentraceae do not have them. Vessels are also absent in some parasitic, xerophytic and aquatic genera where they have been lost through environmental adaptation. The size, number and position of the vessels in the yearly growth rings are important as is the position of the parenchyma. Longitudinal sections can then reveal the nature of the pitting in the conducting cells, which will help in deciding if there are vessels and what their pits are like or if there are only tracheids. The characters of the rays are best examined in TLS if there is only the opportunity for one vertical cut.

Identification of fossil woods is achieved through comparing them with wood of extant species. There are a number of books on the identification of woods that have been produced for the timber trade including Hoadley (1980) and Wheeler *et al.* (1989).

these two types of wood). As trees grow it is only their outer wood that is conductive (sapwood) while the older central wood (heartwood) becomes non-functional and the parenchyma dies. The cells are usually often impregnated with metabolic waste products such as tannins, oils, resins and colouring compounds. Vessels are often blocked with balloon-like outgrowths (tyloses) from surrounding parenchyma through their pits. For this reason heartwood is the more valuable timber. Pits are formed during the development of the cell walls. The primary wall is thinner and differs in composition in the areas where pits are forming. These primary areas, called 'membranes', that are left uncovered when the rest of the primary wall is covered by secondary thickening are called pits. Most pits are circular, but some may be elongated. To be functional both walls of contiguous cells must have a coincident pit, hence the term pit-pair. Occasionally there are 'blind-pits' where there is a pit only in one wall and 'unilateral

compound-pits' where one pit is contiguous with two or more smaller pits in the other wall. Pits may be simple, with vertical edges, or bordered where the secondary wall thickening forms an open dome over the membrane. If only one side of the wall is bordered the pit is said to be half-bordered. Sometimes the central portion of the membrane develops to the same extent as the rest of the primary wall to form what is known as a torus. The flexible membrane in the living spring wood permits the torus to change position with changes in turgor pressure and in adverse conditions can be forced against one aperture, effectively controlling the passage of water.

Ancestors of the angiosperms

For many years the origin of the flowering plants was thought of as a complete mystery, and ideas of their ancestors were based upon anatomical

Table 10.2	*Characters that distinguish angiosperm from conifer wood.*	
	Angiosperms	Conifers
Conducting cells	Nearly all have vessels that have a variety of types of pits (sometimes bordered pits with raised incurved edges) in their vertical walls. They connect vertically through inclined end walls (called perforation plates) with scalariform pits (ladder-like). Primitive vessels are long and thin with oblique perforation plates. The most advanced are short and broad with horizontal perforation plates.	Tracheids are both supporting and conducting cells. They are elongated with tapered ends and conduct water between each other entirely through bordered pits in their side walls.
Other vertical cells	Fibre tracheids with slightly bordered pits for support. Fibres of various kinds. Xylem parenchyma in nearly all. Resin canals only very occasionally present.	Fibre-tracheids (differing from tracheids only in wall thickness and size and number of pits). Fibres are rare. Wood parenchyma (resin cells) often present (absent in some e.g. *Araucaria*).
Rays	May be one cell wide (uniseriate), two cells wide (biseriate) or few to many cells wide (multiseriate) but they all taper towards top and bottom to become uniseriate. Usually consist entirely of parenchyma cells but may have ray tracheids (similar to ray parenchyma cells but with very thick walls).	Usually uniseriate. May have ray tracheids.
Tension (reaction) extra thickness of wood formed in response to mechanical stress	Upper surface of branches.	Lower surface of branches.

comparisons with living and fossil gymnosperms. However, in order to progress, the origin and evolution of the angiosperms must be treated as a problem waiting to be solved, rather than as a mystery, and to do this we must look at the available evidence.

Flowering plants, like any complex large plant, are preserved as fragments in the fossil record. Of these, leaves and pollen grains are the most common, with anatomically preserved stems and then flowers being much less frequently found (Fig. 10.4). As outlined above, the major problem is how to be certain that such isolated fossil organs are unequivocally angiosperm in origin, especially when seeking features which might be taken as indicators of ancestral angiosperms or angiosperm precursors. More than twenty Mesozoic taxa including the Gnetales, glossopterids, caytonias, cycads, bennettites and even ferns have been seriously considered to be possible ancestors of the angiosperms, showing how difficult their interpretation can be. The evolution of the angiosperm flower is itself a matter for debate because there is no clear transitional series between gymnosperm reproductive structures and the angiosperm flower. A number of complex hypotheses have been developed to explain how it could have happened, which are usually based on possible homologies between the position and

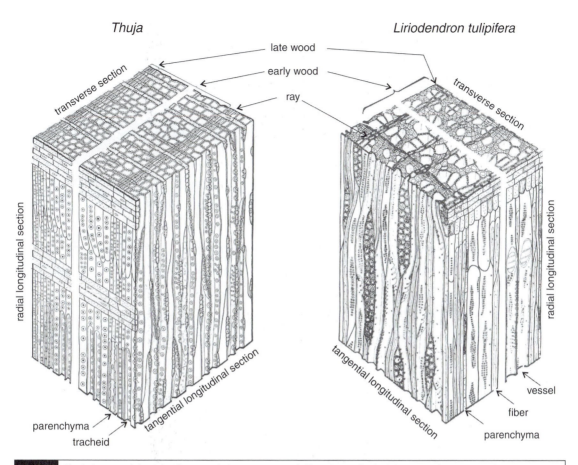

Fig. 10.3 Block diagram of the secondary wood of an angiosperm (left) and a conifer (right) showing faces cut in transverse section (TS), radial longitudinal section (RLS) and tangential longitudinal section (TLS). The secondary xylem consists of vertically orientated conductive (and possibly non-conductive) cells and horizontal rays of parenchyma of varying heights and thickness running through the wood. The rays appear in TS as files of cells running radially between the vessels and tracheids, in RLS as blocks of cells running across the vessels and tracheids, and in TLS they are cut across so they appear as isolated patches of cells arranged in vertical files. In autumn the wood cells produced from the cambium are smaller than those produced in the spring. This gives the distinctive feature known as annual growth rings whereby the age of a stem can be gauged by counting the number of rings. The thicknesses of the annual rings reflect the varying growing conditions to which the tree is subjected. Block diagrams are modified from Esau (1960).

morphology of organs making up the reproductive structures. As yet the matter has not been satisfactorily resolved.

The angiosperms must obviously be descendants of a group that first evolved the critical combination of angiosperm characters outlined earlier. These characters then gave a competitive edge to the plants, enabling them to out-compete their less angiosperm-like contemporaries. Evolution from the earliest Mesozoic groups could be taken to imply that the flowering plant line was distinct by late Triassic times and some DNA sequencing evidence suggests that its origin was even earlier

and possibly in the Carboniferous Period. Fossil evidence based on adpressions and pollen suggests that the main period of angiosperm evolutionary radiation was in the Cretaceous Period, again suggesting that their ancestors must have been much earlier. Proponents of earlier dates have postulated that angiosperms evolved in upland areas that were far from sites of possible fossilisation. However, a subsequent widespread and successful invasion of gymnosperm-dominated lowland floras by such cryptic upland plants does not seem a very likely scenario. It should be made clear that we are not even certain whether the angiosperms

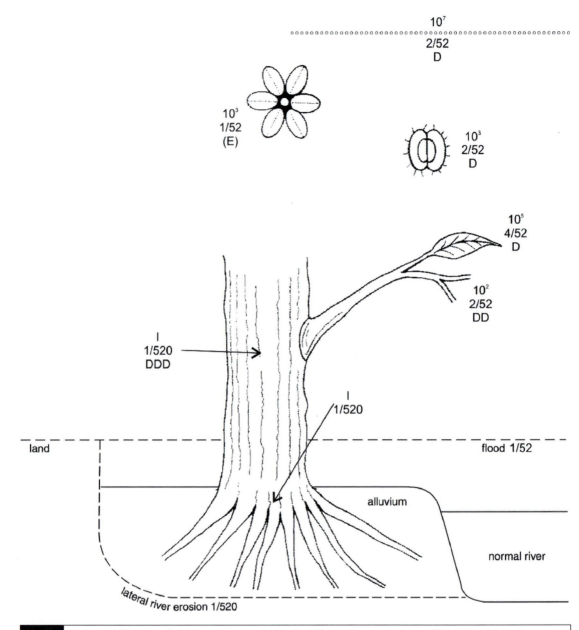

10^7
2/52
D

10^3
1/52
(E)

10^3
2/52
D

10^5
4/52
D

10^2
2/52
DD

I
1/520
DDD

I
1/520

land

flood 1/52

alluvium

normal river

lateral river erosion 1/520

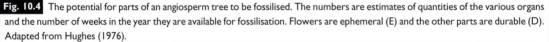

Fig. 10.4 The potential for parts of an angiosperm tree to be fossilised. The numbers are estimates of quantities of the various organs and the number of weeks in the year they are available for fossilisation. Flowers are ephemeral (E) and the other parts are durable (D). Adapted from Hughes (1976).

are really a single group or have their origin in more than one evolutionary event. If the 'angiosperm threshold' was passed simultaneously by more than one group, subsequent evolutionary change and convergence would have obscured the original differences.

The earliest angiosperms

The earliest example of a fossil angiosperm that is most widely cited is a Triassic genus of palm-like leaves called *Sanmiguelia* which was originally described from Colorado, USA (Fig. 10.5a, b).

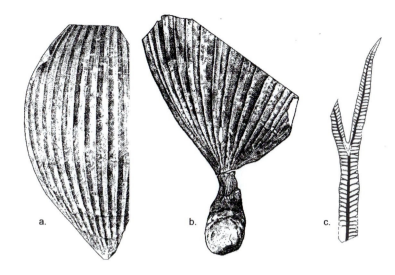

Box 10.2 | Identifying angiosperm leaves

Angiosperm leaves typically consist of a thin, flat blade (lamina) that is supported by a network of veins, a stalk (petiole) and a base where the leaf is connected to the stem. Leaves are called simple when the lamina is a single unit or compound when the lamina is divided into leaflets. Compound leaves may be palmate when the leaflets arise from the same point or pinnate when the leaflets arise from either side of the main stalk (rachis). There may be buds on the stem in the axils of leaves (axillary buds) but there are never buds in the axils of leaflets.

The main features of angiosperm leaves are shown in Fig. 10.6. Leaves (or leaflets) usually have a large central vein, producing a slight thickening (midrib), from which arise lateral veins that give rise to the network of veins running throughout the leaf. There is considerable variation in the layout of veins within the angiosperms. A variation that taken together with the overall shape of the leaf, the shape of its apex and base and the type of leaf margin (e.g. smooth, dentate, serrated) allows leaves to be distinguished and classified and comparisons to be made between fossil leaves and leaves of extant families, genera and sometimes even species.

The plants were about 0.6 m tall with helically arranged leaves on a conical woody stem. Reproductive organs, called *Synangispadix*, have been found in the same beds as the leaves. They had numerous sessile microsporophylls helically arranged on an axis with each bearing two elongated pollen sacs. The name *Axelrodia* is given to the ovulate parts of the plants that bore numerous structures thought to be carpels. Isolated seeds called *Nemececkigone* are also found in the same beds. *Sanmniguelia* can be cautiously interpreted as an angiosperm showing features of both monocotyledons and dicotyledons.

Furcula, from the Triassic System of Greenland, is another possible early angiosperm based upon its lanceolate dichotomised leaves with its distinctive venation pattern of many lateral veins coming from a prominent midrib (Fig. 10.5c).

Dispersed pollen grains have also been cautiously referred to the angiosperms. These include *Eucommiidites*, known to range from

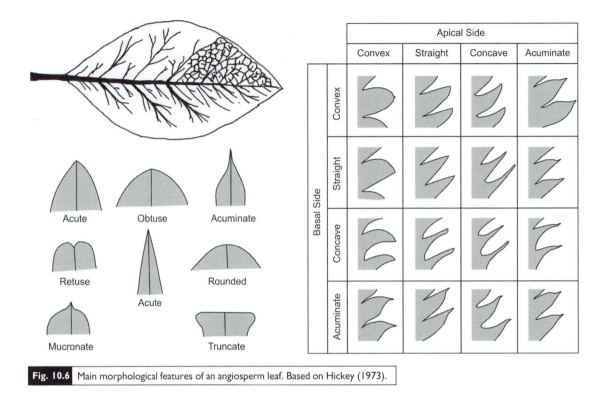

Fig. 10.6 | Main morphological features of an angiosperm leaf. Based on Hickey (1973).

Triassic to Cretaceous in age, and *Equietosporites* and *Cornetipollis* of Triassic age.

Cretaceous angiosperms

It is generally accepted that the earliest flowering plants in the Cretaceous Period were most probably perennial herbs or small shrubs that lived in the more open and disturbed habitats alongside rivers. This was a time of global warmth; the poles were probably free of ice, and the lower latitudes in Early Cretaceous times were dry or seasonally dry. The Early Cretaceous vegetation was divided into several provinces that were dominated by different assemblages of gymnosperms (see Chapter 11). The climatic conditions of the time may have stimulated the evolution of such weedy plants with their photosynthetically efficient planated leaves, reticulate leaf venation (that provided an efficient transport system even when damaged), better vascular systems with vessels, and flowers that enabled quicker and more efficient reproduction than the gymnosperms with their cones. It is interesting that

the Gnetales also arose at this time, because they show many parallels with the flowering plants in having flower-like reproductive organs, vessels in the wood, and in *Gnetum* large veined leaves.

Original ideas of the early evolution and radiation of flowering plants were almost entirely derived from studies of Cretaceous fossil leaves. The leaves were closely compared with those of extant families and often referred to extant genera. Newer studies not only show that practically all of the earlier identifications were incorrect but that successively younger Cretaceous flowering plant floras show the increasing levels of complexity that are predicted by many modern classification systems (Fig. 10.7). Early Cretaceous leaves from Central Asia, the Russian Far East, Portugal and eastern USA were small, simple, entire-margined and pinnately veined, and some were irregularly lobed. Middle Cretaceous leaves were quite varied, either pinnately compound or palmately lobed with more regular palmate venation patterns. Others had a low level of vein organisation, while one *Magnolia*-like form had the same vein pattern seen in the living members of the

ZONES

Upper
2B

Lower
2B

I

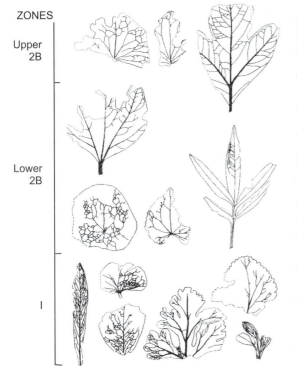

Fig. 10.7 Examples of Mid-Cretaceous angiosperm leaves from the Potomac Group of North America showing increasing levels of overall leaf morphology and vein organisation in successive higher levels. The leaves in Zone I are generally small with disorganised venation and are largely restricted to sandy stream margin lithofacies with the largest possibly being ancestors of woody magnolias adapted to being understory plants. The leaves in the middle 2B Zone are of several new types. Ovate-cordate and peltate leaves may be the ancestors of floating aquatics. The pinnatifid and later compound leaves that show increasingly regular venation were probably pioneering 'weed trees'. The palmate leaves appearing in the upper part of the sequence are best interpreted as belonging to riparian trees that were ancestors of the Hamamelidales. From Thomas & Spicer (1987), after Doyle & Hickey (1976).

Cuticle studies can be important in studying fossil leaves with a view to seeing the arrangement and types of epidermal cells, the organisation of the stomatal complex made up of the guard cells and the surrounding cells, and the surface ornamentation including the presence or absence of hairs, scales and/or glands.

For many years, it was assumed that flowers were too delicate to have become fossilised, but an increasing number of flowers have recently been found and many of them are referable to extant families. The first convincing flower fossils are Late Cretaceous in age (e.g. Fig. 10.8), although Early Cretaceous fragments may subsequently be proved to be of flowers (Fig. 10.9).

Pollen grains dispersed from angiosperm flowers provide an additional means for monitoring change and evolutionary radiation through time. The oldest known pollen grains, of Early Cretaceous age, were simple and of the type restricted to magnoliids and monocotyledonous plants. Soon after this, four other classes of flowering plants can be recognised by their pollen.

Macrofossils were thought to support the idea that the first recognisable flowering plants had small and simple flowers of the *Magnolia*-type with no differentiation of petals and their surrounding green sepals. The pollen producing anthers and the ovule-bearing carpels were simple and relatively few in number. Interestingly, the first *Magnolia*-like leaves are rare, even when they are at their most abundant in gymnosperm and fern dominated plant fossil assemblages. This suggests that these plants grew in the forest as a shrub layer. The mid-Cretaceous *Lesqueria* from Kansas is such an example of a *Magnolia*-type of flower in having a receptacle with numerous carpels containing about 10–20 seeds (Fig. 10.10c). Another is the Cretaceous fruit called *Archaeanthus* that has up to 300 tightly packed elongated carpels with nearly 100 seeds in each (Fig. 10.10a). Scars on the axis below the carpels are thought to mark the former positions of stamens, perianth and bud scales. Large, deeply lobed leaves associated with *Archaeanthus* are called *Liriophyllum*.

There were, however, earlier examples of flowers found in Lower Cretaceous beds that

Magnoliales. Such close similarities must be treated with caution and cannot be taken as unequivocal evidence of a close taxonomic relationship. For example, there were roughly circular leaves with radiating major veins that can be closely compared with those of some living climbers and aquatics. Another major group had palmate veins and sometimes toothed margins, typified by the leaves of the *Araliosoides*, *Sassafras*, and the *Platanus* types.

Fig. 10.8 Small carbonaceous three-dimensionally preserved flowers of Late Cretaceous age from Sweden and Portugal. All scale bars = 0.2 mm. a, b. *Silvianthecum suecicum* Friis, which can be closely compared to the woody saxifragalean family Escalloniaceae, now growing predominantly in the southern hemisphere. c. An unnamed flower with an inferior ovary, perianth lobes, pollen sacs and a central style. d. Single stamen of *Platanus scanius* Friis et al. e, f. *Esgueiria adenocarpa* Friis et al., which is most closely related to the West African genus *Gueira* in the Combreteae. The flowers were borne in dense heads although their arrangement is not known. The illustrated flowers show five sepals, but the petals have been lost. g, h. A single unnamed flower consisting entirely of stamens. Its systematic affinity is unknown although the structure of its pollen suggests it to be a magnoliid. All the specimens are in the collections of the Swedish Museum of Natural History, Stockholm. Photos by E. M. Friis (from Cleal & Thomas, 1999).

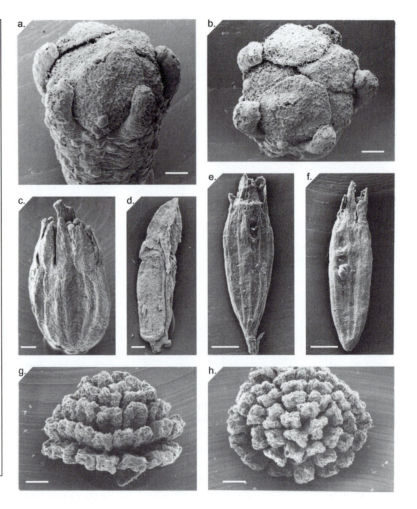

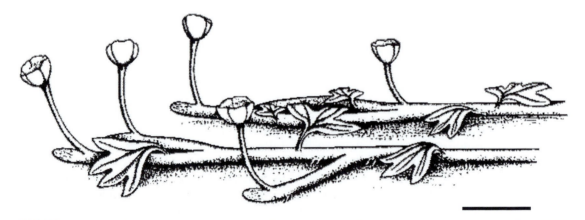

Fig. 10.9 A reconstruction of *Bevhalstia*. Drawing by A. Townsend, based on the work of Jarzembowski in Hill (1996)

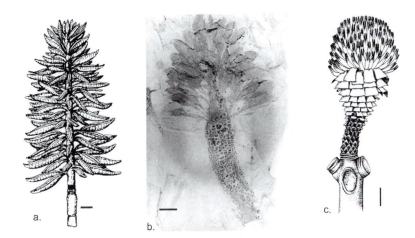

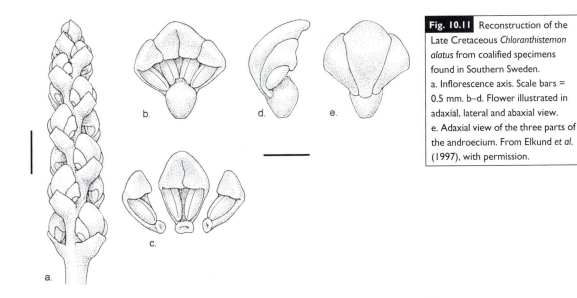

Fig. 10.10 Cretaceous fruiting structures from Kansas, USA. a. *Archaeanthus* marking the original position of the other flower parts. Scale bar = 0.1 mm. Drawing from Thomas & Spicer (1987). b. An unnamed angiosperm fruit from the Dakota Formation (Middle Cretaceous) of central Kansas, USA (Florida Museum of Natural History, Gainesville, Specimen 15706–3084). Photo by D. Dilcher. c. A fruiting axis of a primitive *Magnolia*-like angiosperm, *Lesqueria*, from the Middle Cretaceous Series. The carpels are borne helically in an ovoid head showing it to be most similar to Recent magnoliid angiosperms. Scale bar = 10 mm. Drawing from Cleal & Thomas (1999).

Fig. 10.11 Reconstruction of the Late Cretaceous *Chloranthistemon alatus* from coalified specimens found in Southern Sweden. a. Inflorescence axis. Scale bars = 0.5 mm. b–d. Flower illustrated in adaxial, lateral and abaxial view. e. Adaxial view of the three parts of the androecium. From Elkund *et al.* (1997), with permission.

are tentatively referable to the Chloranthaceae (e.g. *Chloranthistemon* from southern Sweden – Fig. 10.11) and Platanaceae (e.g. *Platananthus* from North America and Sweden).

A considerable amount of information now exists about the evolution and diversification of angiosperms through the Mesozoic Era and the effect that they had on the vegetation as a whole. By Middle Cretaceous times, angiosperms were starting to dominate some parts of the vegetation. There were freshwater swamp woodlands dominated by plants bearing *Magnoliophyllum*, *Liriophyllum* and *Sapimdopsis* leaves, platanoid-leaved plants were abundant on channel levees

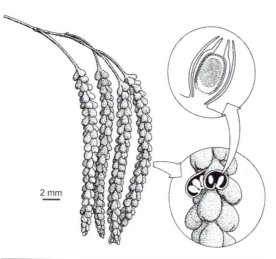

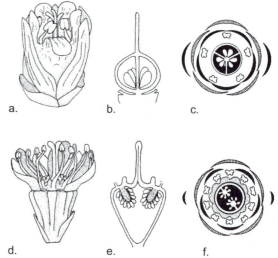

Fig. 10.12 Reconstruction of the Mid-Cretaceous *Prisca reynoldsii* Retallack & Dilcher from Kansas, USA, showing a simple catkin-like raceme of elongated follicles, together with a cut-away of the follicles and a longitudinal section of a single seed showing outer and inner integuments enclosing the nucellus with its large megaspore (dashed line = megaspore membrane). From Retallack & Dilcher (1981).

Fig. 10.13 Organisation and arrangement of parts in two radially symmetrical Late Cretaceous flowers from Sweden showing a reconstruction of the flowers, sections through their ovaries to show placentation and floral diagrams. a–c. *Actinocalyx* with five free sepals, five partially fused petals, five stamens fused basally to the petals, a trilocular superior ovary with three styles. These have possible affinities with the Ericales and Ebenales. d–f. *Scandianthus* with five free sepals, and five free petals (lost after pollination), ten stamens in two whorls, and an inferior unilocular ovary formed from two fused carpels. These may be related to the saxifragalean complex. From Friis (1985), with permission.

and around lakes, and there were mangrove-like communities on the tidal deltas.

The time was clearly right for the flowering plants, because there was now a massive evolutionary burst, producing many other families before the end of the Cretaceous Period. Apart from the magnoliid forms, there were now primitive hamamelids (witch hazels), laurels, planes and others with buttercup-like flowers. Other well-differentiated flowers in the fossil record also provide evidence of a number of 'experimental' families that are now extinct. Some were very different from any living angiosperms. For example, the Mid Cretaceous angiosperm fructification *Prisca reynoldsii* Retallack & Dilcher (Fig. 10.12) is attributed to the same extinct plant as the leaves *Magnoliaephyllum*, suggesting it was a woody shrub or tree with such an individual combination of characters that it should be in its own family, the Priscaceae.

This floral radiation was greater than at any previous or subsequent time. It can be related to changing climate and plate tectonics, which led to tropical conditions with rain forests that were unparalleled since the early Permian Period. Many flowers were now more specialised

and some, like saxifrages (e.g. Fig. 10.13; see Fig. 10.14 for a modern-day saxifrage flower) had well-developed nectaries to attract Hymenoptera and stingless honeybees. The low level of specialisation in floral morphology, coupled with indiscriminate pollen transfer and the probability of genetic compatibility between such early members of radiating groups, probably led to high levels of interbreeding. This would have accelerated morphological variability and new character combinations that might have enabled the hybrids to out-compete their parental populations, possibly leading to population segregation. Isolation could then have led to further genetic and morphological adaptation resulting in better defined species boundaries.

By the end of the Cretaceous Period, 31 extant families of flowering plants are clearly recognisable in the fossil record including the Magnoliaceae

Many early flowering plants must have been an attractive source of food with their leaves, flowers and fruits being more easily digestible than conifer shoots and cones. They were small and easily accessible to browsing, but could withstand its effects by being able to regenerate themselves better than any of the gymnosperms. This must have been especially important for young saplings, which would certainly have given the flowering plants a great competitive advantage over all the conifers.

The mid-Cretaceous diversification of angiosperms is accompanied by the transition from a sauropod-dominated dinosaur fauna to one dominated by the browsing ornithopods which had efficient grinding teeth. The larger food supply provided by the angiosperms led to the herbivorous dinosaurs increasing in number and diversity. The larger flesh eaters, such as the theropods *Tyrannosaurus* and *Gigantosaurus*, then evolved to prey on them, resulting in a food chain similar to those seen today in mammals.

There is some evidence of leaf herbivory as far back as the Carboniferous Period and by the Jurassic Period there is clear evidence of leaf-miner damage in the gymnosperm *Pachypteris*. But the flowering plants with their broad, comparatively fleshy leaves were a much more attractive source of food than conifers. Within 25 million years of early flowering plant radiation, insect herbivores were exploiting them as their major food source. Flowering plants were able to survive this constant onslaught through their ability to shed their older leaves and replace them with new ones.

Some flowering plant seeds can survive passage through the digestive tracts of the animal browsers, which greatly aids their dispersal. The evolution of succulent seed coats and the swelling of ovaries to give fruits added to their attractiveness as food and improved their potential dispersal.

The other plant/animal interaction that the rise of the flowering plants produced was pollination by insects. For effective reproduction, pollen must germinate on the receptive surfaces of the female reproductive organs. The means by which

Fig. 10.14 *Saxifraga stellaris* L. growing in Ben Lawers National Nature Reserve, Perthshire, Scotland. Photo from NMW Collection.

(magnolias), Hamamelidaceae (witch hazels), Juglandaceae (walnuts), Caryophyllaceae (pinks), Leguminosae (peas) and the Araliaceae (ivies). Angiosperms abscise their older leaves as they grow, so a further adaptation to periodical drought during their early evolutionary radiation could have permitted some species to shed all their leaves synchronously to precede a dormant phase. This deciduous habit would have allowed angiosperms to colonise areas that were subjected to periodic water or temperature stress where the gymnosperms could not survive. It also permitted them to spread to higher altitudes and palaeolatitudes where lower winter light and temperatures could be survived in a dormant condition. There is evidence that many angiosperms migrated northwards during late Cretaceous times, most probably along the coastal plain. Once north of the closed conifer forests they would have spread away from the coast into the interior.

pollen is transferred is varied, but in living angiosperms insects are among the commonest vectors, and complex protein-recognition processes usually ensure that only pollen of the right type germinates to effect nuclear fusion. Although insect pollination occurred in some gymnosperm groups, it was developed to its highest level in the angiosperms. Such plant-pollination relationships reward the pollinator for visiting the plant while they incidentally effect pollination. The earliest reward was the pollen itself, but the development of nectar, together with more organised and specialised flower parts, led to many different co-evolutionary paths of flowers and their pollinators.

Wind pollinated angiosperms

Not all flowers were insect pollinated. There is a view that non-biological pollination, particularly by wind, was a later adaptation that became important during Cretaceous times, although insect pollination became more dominant again in Palaeogene times (Fig. 10.15). Wind-pollinated flowers are typically unisexual and they may, or may not, be on the same plant. There is not necessarily any consistency in this feature within families or even within a genus. The male flowers are often arranged into catkins, with the earliest well-documented catkin-like structure being of earliest Late Cretaceous age from Kansas, USA. This is *Prisca reynoldsii* mentioned above (Fig. 10.12) which grew in woodland vegetation on waterlogged peat fringing

brackish to freshwater lagoons in a delta plain. *Prisca* produced large numbers of small seeds apparently dispersed mainly by wind and water, and it was most probably wind pollination that made the plant a successful early coloniser in the unstable coastal plains. There are also ball-like seed heads known from roughly the same locality that are very similar to those of extant *Platanus* (planes). There is leaf and pollen evidence of similar age to the wind pollinated mimosids from Kansas although the first evidence of their inflorescences is of Eocene age. Leaf evidence suggests that the birches evolved in Cretaceous times but again reproductive structures are not known until the Oligocene Series.

The rise of the monocotyledons

The origin of the monocotyledons is another problem that confronts us. Today, arborescent forms dominate certain taxa such as the palms (Palmae), Pandanaceae and some Agavaceae, and occur in other groups that are mainly herbaceous. Some of the arborescent monocotyledonous genera undertake secondary growth, including *Yucca* (Agavaceae), *Dracaena* (Dracaenaceae), *Cordyline* (Asteliaceae) and *Aloe* (Asphodelaceae) but none of them achieve the dimensions of trees. A range of growth-forms of living monocotyledons is shown in Fig. 10.16. We do not know if such a disparate range really is a monophyletic group or whether the

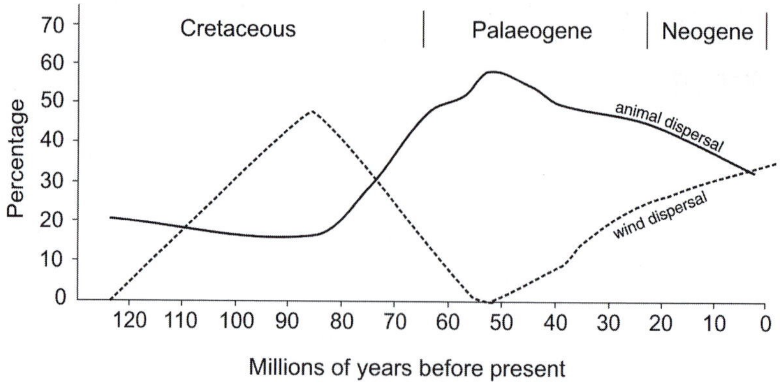

Fig. 10.15 Proportion of animal-dispersed angiosperm taxa and wind-dispersed angiosperm taxa in 25 fossil floras ranging in age from Early Cretaceous to late Cenozoic. Based on information in Eriksson *et al.* (2000).

Fig. 10.16 Living monocotyledon plants. a. *Juncus acutus* L. (spiny rush), Whitford Burrows, England. b. *Cocos nucifera* L. (coconut palm), Ghana. The palms can grow up to 30 m tall and have a crown of pinnate leaves up to 6 m long. They require warmth for successful growth which is greatly reduced at temperatures below 21 °C. Coconut palms have both male and female flowers in the same inflorescence and flowering is continuous with the female flowers producing the large seeds (coconuts). c. *Agave angustifolia* Haw. (Caribbean agave), with its rosettes of fleshy, spiny leaves, central Guatemala. d. *Yucca brevifolia* Torr. (Joshua Tree) lives mainly in the open grasslands of the Mojave Desert in southwestern USA; it grows just over a centimetre a year and can live for several hundred years. Photos by P. Russell (a) and B. A. Thomas (b–d).

Fig. 10.17 a. A palm flower from Dominican Cenozoic amber referable to the extant genus *Trithrinax*. Scale bar = 0.5 mm. From G. Poinar (2002). b. A cut slab of a Cretaceous silicified palm from Austin, Texas, USA. Scale bar = 5 mm. Palms, being monocotyledonous angiosperms, have many individual vascular bundles in their stems and do not form secondary wood, as do conifers and dicotyledonous angiosperms. From Cleal & Thomas (1999). c. A transverse section of *Protoyucca shadishii* Tidwell from the middle Miocene Series of Nevada, USA. The central primary vascular bundles are surrounded by rows of peripheral bundles and then the darker layer of secondary bundles. *Protoyucca* could grow to 0.6 m in diameter. Photo by D. Tidwell. d. Transverse section of the petiole of the monocotyledon *Heleophyton helobiaeoides* Erwin & Stockey showing the outline with lateral ribs (r) and channels (arrowed) and the distribution of the oval vascular bundles in arcs with the central bundle of the median arc (arrowed). Scale bar = 0.2 mm. From Erwin & Stockey (1989), with permission.

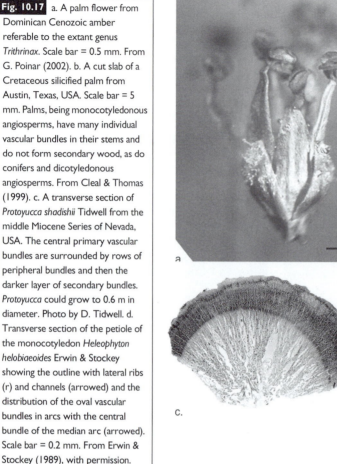

a.

c.

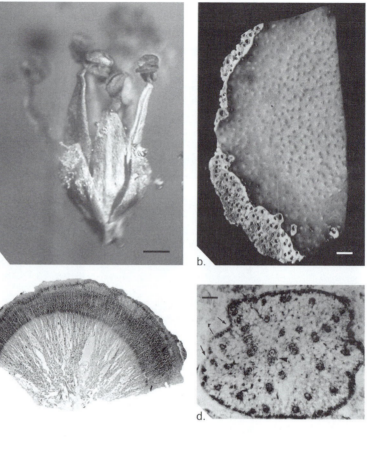

b.

d.

constituent families had separate origins from the dicotyledons.

The earliest unequivocal monocotyledons are of Late Cretaceous age, such as palm wood, known as *Palmoxylon* from Texas, USA (Fig. 10.17a). Other palm fossil stems are known from the Eocene Series of south central USA and the Eocene and Oligocene series of central Europe; palm roots from the upper Eocene Series of Utah, USA; and Late Cretaceous and Cenozoic palm leaves are known from the southern USA.

Protoyucca shadishii (Fig. 10.17c) from the Middle Miocene of Nevada, USA, is a permineralised monocotyledon with secondary growth. Associated leaves and swollen bases with roots are referred to this species showing it to be *Yucca*-like in appearance and in anatomy. It was probably a subtropical ancestor of *Yucca*, living in dry edaphic sites.

There are Cenozoic fossils from the USA that are closely comparable to extant members of the Araceae, such as leaves included in *Philodendron* and florets similar to those in *Acorus*. Other semi-aquatic monocotyledons from the Eocene beds of British Columbia, called *Heleophyton* (Fig. 10.17d), belong to the Alismataceae. The Cenozoic Era also saw the rise of grass (Fig. 10.20) and the grasslands that had such a major impact on the

Fig. 10.18 Cenozoic angiosperms of the northern hemisphere. Scale bars = 10 mm. a. *Cedrelospermum lineatum* (Lesquereux) Manchester from the early Oligocene Florissant Beds of central Colorado, USA, can be referred to the extant family of elms (Ulmaceae). The twigs have petiolate, asymmetrical, lanceolate leaves with acute apices and bases and a toothed margin. The attached fruit is a small winged seed, a samara attached by a short stalk. Peabody Museum, Yale University (Specimen 25232). Photo by S. R. Manchester. b. *Polyptera manningii* Manchester & Dilcher (Juglandaceae) from the Palaeocene Series of southern Wyoming. Several fruits with a single wing divided into lobes surrounding a rounded-triangular nut tapering from its broad base to a pointed apex. These are consistently associated compound leaves *Junglandiphyllites glabra* (Walt) Manchester & Dilcher. Florida Museum of Natural History, Gainesville (Specimen 13687). c. *Paracarpinus chaneyi* (Lesquereix) Manchester & Crane (Betulaceae) from Oligocene Beds in Oregon, USA, showing uniformly spaced straight secondary veins, toothed margin and intact petiole. Florida Museum of Natural History, Gainesville (Specimen 6096). d. A sycamore (*Acer* sp.) leaf from Pleistocene diatomaceous deposits (Pretiglien-Tieglien) near Faufouille, France (NMW Specimen 86.54G.1a). From Cleal & Thomas (1999).

Earth's terrestrial biotas; these will be discussed in the next section.

Cenozoic angiosperms

Angiosperms appear to have been little affected by the Cretaceous-Tertiary (K/T) extinction event that caused so much havoc with animal life. During the Cenozoic Era, angiosperms continued to diversify rapidly, eventually dominating most of the world's land vegetation. Our knowledge of angiosperm evolutionary history is still far from complete, with only about a half of the 300–400 living families having been identified in the fossil record. Of these, however, just under a hundred first evolved during the Palaeogene Period. Especially in the northern hemisphere, the Salicaceae, Ulmaceae, Betulaceae and Juglandaceae (Figs. 10.18, 10.19) became

Fig. 10.19 The Middle Eocene *Populus wilmattae* Cockrell from northwestern Utah, USA, shows leaves and fruits attached to the same twig (Palaeobotanical Collections, Brigham Young University, Utah, Specimen 3003). Scale bar = 10 mm. The leaves are ovate to deltoid with acute to long-attenuate apices, toothed margins and long petioles. The fruiting axis is a raceme that has a potentially continuously growing apex with the oldest parts at the base, with ovate three-valved fruits (capsules). Photo by S. R. Manchester. From Cleal & Thomas (1999).

major elements in the flora. A key factor in explaining this Palaeogene explosion in diversity is the global climatic cooling that probably started in middle Eocene times. This would have increased the stress on vegetation, which had become largely adapted to the warmer conditions in Mesozoic and early Palaeogene times, and would have provided an opportunity for diversification in the more adaptable groups of plants, such as the angiosperms.

This climate change continued through the Neogene Period, culminating eventually in the Quaternary ice age. The most important Neogene development in land vegetation was the expansion of open grasslands. Today there are more than 8000 species of grasses but the fossil record is very sparse. Some Late Cretaceous pollen might be from grasses, but the earliest generally accepted records of grass pollen are of early Palaeocene age, from Brazil, Africa and Australia. The earliest fossilised leaves are also of Palaeocene age, but there are increasing numbers of fossil grasses in the Eocene when many modern families of grasses appeared. Some of these grasses, like the Late Miocene *Tomlinsonia* from California have anatomical features, such as prominent bundle sheaths, close veins and radiating mesophyll, which

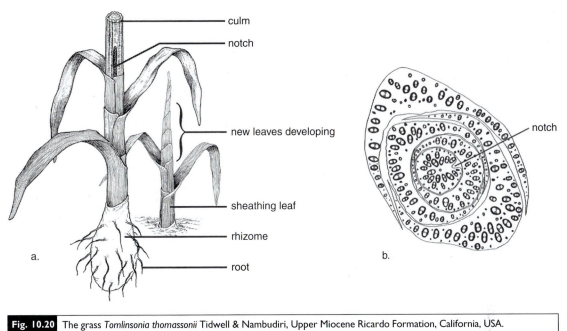

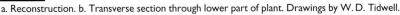

Fig. 10.20 The grass *Tomlinsonia thomassonii* Tidwell & Nambudiri, Upper Miocene Ricardo Formation, California, USA. a. Reconstruction. b. Transverse section through lower part of plant. Drawings by W. D. Tidwell.

suggests they were C_4 plants (Fig. 10.20). C_4 plants can photosynthesise faster than most other plants (C3 plants) under high light intensity; this allows them to have their stomata closed for much of the time, thus reducing water-loss through transpiration. Bamboos are thought to have evolved separately from the grasses but the two have been found together in the Miocene Series suggesting that they may have coexisted. The vegetational change caused by the proliferation of grassland had a profound impact on animal life, providing the trigger for the evolution of the herbivorous ungulates that have become among the most numerous mammal groups. It also provided the ancestral forms of the cereal and forage crops on which people are now so dependent.

As in the Mesozoic Era, much of our knowledge of Cenozoic angiosperms is derived from leaf fossils (e.g. Fig. 10.21). Angiosperm foliage studied in isolation from the rest of the plant can be very difficult to interpret but considerable progress has been made in recent years through the detailed analysis of venation patterns and leaf-margin characters. Some flowers have been found (e.g. Fig. 10.22) but they are generally rare. More abundant are fossil fruits and seeds, and these tend to be a reliable means of identifying angiosperm families. There are a number of classic Palaeogene fruit and seed floras, such as from the London Clay of southern England (Fig. 10.23) and the Clarno Nut Bed of Oregon (Fig. 10.24), and these have provided a clear insight into the rapid diversification that was taking place in angiosperms during Palaeogene times.

Isolated organs such as fruits and seeds are often very similar to those of some living

Fig. 10.21 Cenozoic angiosperm leaves referred to extant genera. All scale bars = 5 mm. a. *Quercus pseudocastanea* Göppert. From the Neogene Series of Europe. From van der Burgh (1993) with permission. b. *Acer tricuspidatum* Bronn. c. *Zelkovia zelkovifolia* (Ung.) Búřek & Kotlaba. d. *Platanus neptuni* Búřek et al. e. *Engelhardia orsbergensis* (Wessel & Wessel) Jähnichen et al. f. *Magnolia* sp. From the Oligocene Series of the Czech Republic (b–f). From Kvaček & Walther (1995), with permission.

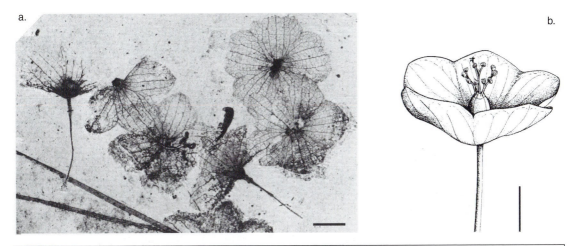

Fig. 10.22 *Florissantia quilchenensis* (Mathews & Brooke) Manchester of Middle Eocene age from Washington State, USA. These plants are thought to have been related to the extant Malvales. Both scale bars = 5 mm. a. Photograph of fossil. Stonerose Interpretive Centre, Republic, Washington State, USA (Specimen 87–26–4). Photo S. R. Manchester (Gainesville). b. Reconstruction of flower. Drawing by A. Townsend, based on the work of S. R. Manchester.

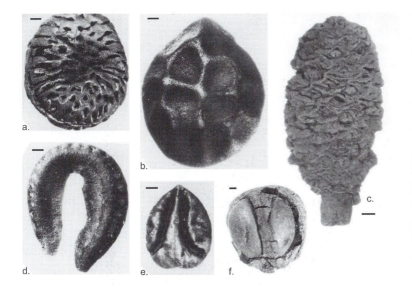

Fig. 10.23 Fruits and seeds from the Eocene London Clay of southern England. Scale bars = 0.5 mm, except in c, where it is 2 mm. a. *Anonaspermum cerebellatum* Reid & Chandler (Anonaceae). b. *Iodes corniculata* Reid & Chandler (Icacinaceae). c. *Platycarya richardsonii* (Bowerbank) Chandler (Juglandaceae), compound fruit. d. *Diploclisia auriformis* (Hollick) Manchester (Menispermaceae). e. *Parthenocissus monasteriensis* (Reid & Chandler) Chandler (Vitaceae). f. *Leucopogon quadrilocularis* Reid & Chandler (Epacridaceae). From Sheppey (a, b, d, e), Herne Bay (c) and Bognor Regis (f). Photos by M. E. Collinson. From Cleal *et al.* (2001).

Fig. 10.24 a–e. Fruit and nuts from the Middle Eocene Clarno Nut Beds flora, Oregon, USA. Scale bars = 1 mm, except in f, where it is 5 mm. Florida Museum of Natural History, Gainesville, Specimens 5236, 6533, 6556, 6621, 9859. a. *Vitis tiffneyi* Manchester (Vitaceae) showing basal view of a seed cast with much of the seed coat flaked away to reveal the internal mould of the seed cavity. The circular mark shows the position of original attachment. b. *Magnolia muldoonae* Manchester (Magnoliaceae) preserved as a chalcedony cast. c. *Eneste oregonense* Manchester and Kress (Musaceae). d. *Saxifragispermum tetragonalis* Manchester (Flacourtiaceae) fruit consisting of an elongate, four-valved capsule with rows of ovoid seeds. The specimen illustrated here has part of the fruit wall missing to reveal the locule of the capsule. e. *Deviacer wolfei* Manchester (Sapindales). These seeds resemble the samaras of living *Acer* (sycamores) although there are several subtle differences that are sufficient to separate them. f. *Corylus avellana* L., from Recent post-glacial deposits overlying Coed Madog Quarry, Nantlle, Caernarfonshire, Wales. NMW Specimen 27.110G.1337. Photos by S. R. Manchester (a-e) and NMW Photography Department (f). From Cleal & Thomas (1999).

plants, and in some instances Palaeogene fossils have been assigned to living genera, suggesting that relatively little evolutionary change had taken place over the intervening 50 million or so years. For instance, large *Nypa* fruit have been widely reported from Eocene floras of Europe, and these appear very similar to those of the living mangrove palm (Fig. 10.25). Recently, though, it has been realised that a proper understanding of angiosperm evolution can only be obtained by studying at least partly reconstructed plants. From such reconstructions it appears that, although individual plant organs may seem to

Fig. 10.25 a. *Nypa burtinii* (Brongniart) Ettingshausen pyritised fruit from the Eocene London Clay on the foreshore of the Bracklesham Bay, Hampshire, England. Scale bar = 20 mm. b. Extant *Nypa*, seen here in Brunei Darussalam, is a stemless palm, which fringes brackish water courses in South East Asia. c. Illustration (reduced) of *Nypa* fruit from Sheppey, Kent, England, from the original description of the flora by Bowerbank (1840). For a field guide to the London Clay flora see Collinson (1983). d. Closer view of Nypa mangrove, Kapuas Delta, Kalimantan, Indonesia. Photos by M. E. Collinson (a, d) and B. A. Thomas (b).

have changed little, Palaeogene angiosperms show different combinations of organs. For instance, one Palaeogene species from southern England had a winged nut similar to that of the living hornbeam, while the arrangement of the bracts is more similar to that of the hazel, and the foliage is different again from that found on either plant today (Fig. 10.26). It is becoming clear that considerable evolutionary change has taken place among the angiosperms during the Cenozoic Era; a fact that should not come as a total surprise in view of their evident adaptability and success in today's vegetation.

Fig. 10.26 *Palaeocarpinus* plants reconstructed by Crane (1982) from Palaeogene fossils from Cold Ash Quarry, southern England. a. Reconstruction of shoot with leaves and fruits. b. *Palaeocarpinus laciniata* Crane fruit. Scale bar = 10 mm. c. *Craspedodromophyllum acutum* Crane leaf. Scale bar = 5 mm. Photos by P. Crane (b, c). From Cleal *et al.* (2001).

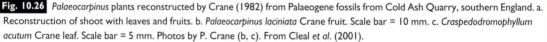

Recommended reading

Beck (1976), Brack-Hanes & Greco (1988), van der Burgh (1993), Call & Dilcher (1992), Benton (1993), Collinson (1983), Crane (1982, 1989), Crane & Blackmore (1989), Crane & Dilcher (1985), Crawley (1989), Crepet *et al.* (1991), Dilcher (1973), Eriksson *et al.* (2000), Erwin & Stockey (1991), Friis (1985), Friis & Endress (1990), Friis & Skarby (1982), Friis *et al.* (1987, 1994, 1997), Herendeen & Dilcher (1992), Herendeen *et al.* (1990), Hughes (1994), Kvaček & Walther (1995), Labandiera *et al.* (1994), Manchester (1987, 1992), Manchester & Crane (1983), Retallack & Dilcher (1981), Roselt *et al.* (1982), Stockey (1987), Taylor & Hickey (1996), Tidwell & Nambudiri (1989), Tidwell & Parker (1990).

Chapter 11

The history of land vegetation

So far in this book, we have looked at the fossil record of the major groups of plants. In this last chapter, we will take a broader look at the plant fossil record, to show how land vegetation as a whole has changed through time – when and where the major developments in the evolutionary history of plants took place, and what the major types of vegetation were at any particular time. Dates for the periods are given in millions of years before present (Ma). Readers who are not familiar with the basics of stratigraphy and its nomenclature should consult one of the various standard textbooks on the subject, such as that by Doyle *et al.* (2001).

Silurian Period (416–443 Ma)

Although the oldest macrofossil evidence of land plants is of middle Silurian (Wenlock) age, palynology indicates that they probably first appeared in late Ordovician times. Throughout Silurian times, only small rhyniophytes (or rhyniophytoids) and lycophytes are known (Chapters 4 and 5). Remarkably, even in these early phases of their evolution, plants had an almost worldwide distribution with records from North and South America, Europe, Africa, Central Asia, China and Australia, presumably reflecting the wide dispersal potential of their spores and the lack of any competition. The global composition of these floras also appears to be fairly uniform, although this may in part be due to the problems of differentiating biological species within the plexus of these morphologically very simple plants.

Devonian Period (359–416 Ma)

From its primitive rhyniophyte and lycophyte precursors, land vegetation rapidly diversified during Devonian times. By the end of the period, all of the major divisions of vascular plants except the flowering plants had appeared, with the first record of ferns (or at least their pre-fern ancestors) being in Middle Devonian times, and of sphenophytes and gymnosperms in Late Devonian times (Fig. 11.1).

The Devonian Period also saw the development of several structures that made the vascular plants better adapted to life on land (Fig. 4.5). Secondary wood is the most widely adopted means by which plants can significantly increase their stature and it is first seen in the fossil record in the Middle Devonian Series. The resulting trees (and therefore presumably forests) must have had a dramatic impact on the Devonian landscape. Photosynthetic efficiency was also enhanced by the development of planated leaves in Devonian times. The process seems to have been gradual through Devonian times, developing independently in the sphenophytes, ferns and progymnosperms.

The main reproductive novelty to appear in vascular plants during the Devonian Period was the seed. Although this was again a gradual evolutionary process via heterospory (see Chapter 8), the earliest structures that can properly be called seeds are of Late Devonian age. The seed liberated plants from the necessity of having external moisture to achieve fertilisation and thus allowed them to occupy a much wider

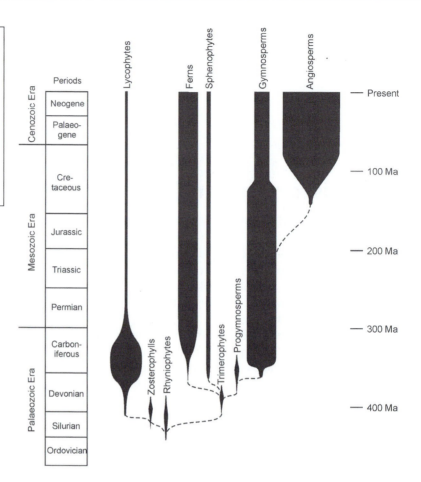

Fig. 11.1 Stratigraphical ranges of the main divisions/phyla of the Plant Kingdom. The dashed lines joining the divisions are intended to show general relationships rather than representing detailed phylogenetic histories. The width of the balloons is intended to give a general impression of relative abundances, but is purely schematic. The Neogene Period incorporates the Quaternary.

range of habitats. It also provided the young plants with a source of nutrition, giving them a great advantage over free-sporing plants, whose prothalli required wet, or at least damp, conditions for growth and fertilisation. Seeds also have the potential to remain dormant for a period of time between fertilisation and germination, allowing them to take advantage of favourable conditions whenever they occur.

It is difficult to assess vegetational diversity and differences in plant distribution in the Devonian Period. Most of our knowledge comes from Europe and North America; our knowledge from elsewhere is still quite poor (one of the best-studied Early Devonian landscapes showing relatively early phases in the development of land vegetation is that of Rhynie – Fig. 11.2). The traditional view was that the global difference in floras was relatively low, especially in early and late Devonian times, with some floral diversity

recognisable in Middle Devonian times. However, new discoveries in China and South America are revealing a much greater biogeographical diversity in Devonian vegetation than previously realised. Nevertheless, we are still a long way from establishing even the broad patterns in the distribution of Devonian plants.

Carboniferous Period (299–359 Ma)

The morphological developments that took place in plants during the Devonian Period provided the springboard from which vegetation could start to take full advantage of the varied terrestrial habitats during Carboniferous times. There were few morphological innovations during the Carboniferous Period. Planated foliage became more common and fructifications more complex (e.g. the development of cones), but

Fig. 11.2 Reconstruction of the Early Devonian volcanic landscape at Rhynie, Scotland. Painting by A. Townsend, based on the work of N. H. Trewin (Aberdeen).

these were in essence just refinements of the structures first seen in Devonian plants. The conifers and cycads first appeared at this time, but mainly in extra-basinal habitats away from the sites where plant remains were being preserved for later fossilisation, and so they are rarely found as macrofossils. What we do see, however, is the spread of plants over large parts of the Earth to produce a pattern of plant communities that is in many ways similar to today's (although, of course, the individual plants were very different).

In Mississippian times, the vegetation in middle and high palaeolatitudes (Angara and Gondwana Palaeokingdoms, Fig. 11.3) was relatively poor and dominated by lycophytes and progymnosperms. In the low palaeolatitude Amerosinian Palaeokingdom, such as in Europe and North America, a more diverse vegetation was flourishing with pteridosperms (mainly lyginopteridalens), pre-ferns and true ferns, and sphenophytes (archaeocalamites) as well as a greater diversity of lycophytes and progymnosperms. Much of the evidence reflects the plants growing in marginal lowland habitats such as lake margins and river deltas. However, there are also some records of plants growing in volcanic terrains, in which the mineral rich waters have often petrified the plants, preserving part of their internal anatomy. Sites in southern

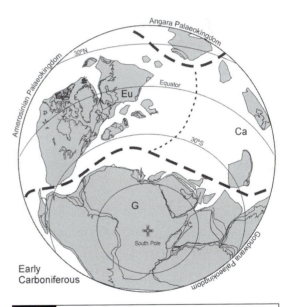

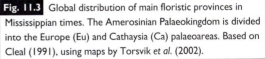

Fig. 11.3 Global distribution of main floristic provinces in Mississippian times. The Amerosinian Palaeokingdom is divided into the Europe (Eu) and Cathaysia (Ca) palaeoareas. Based on Cleal (1991), using maps by Torsvik *et al.* (2002).

Scotland have produced many examples of this type of preservation.

In Pennsylvanian times, there was a dramatic change in vegetation. In low palaeolatitudes, the very first tropical rain forests appeared: the so-called 'Coal Forests' of Europe, North America

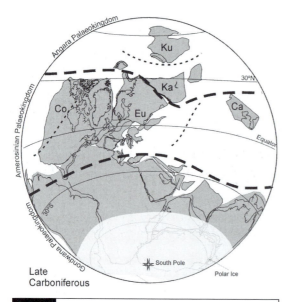

Late
Carboniferous

Fig. 11.4 Global distribution of main floristic areas in Pennsylvanian times. The Amerosinian Palaeokingdom is divided from west to east into Cordillera (Co), Europe (Eu) and Cathaysia (Ca) palaeoareas. The Angara Palaeokingdom is divided into Kazakhstan (Ka) and Kuznetsk (Ku) palaeoareas. Based on Cleal (1991), using maps by Torsvik *et al.* (2002).

and China (known as the Amerosinian Palaeokingdom – Fig. 11.4), so named because of the vast reserves of coal-forming peat that were laid down by them (Fig. 11.5). These forests were dominated by the giant lycophytes (Chapter 5), which were perfectly adapted to the wetland habitats that existed over much of the then tropics. Also within the forests, the somewhat more raised, drier ground (e.g. sand banks and river levees) supported a more diverse range of ferns (including marattialean tree ferns), sphenophytes (calamites and sphenophylls) and pteridosperms (medullosaleans, callistophytaleans, lyginopteridaleans) (Fig. 11.6). Surrounding these wetland forests there were probably conifer-dominated forests, although the remains of these habitats have only occasionally found their way into the fossil record.

At the same time as the tropical lowlands were covered in Coal Forests, much of the southern low-palaeolatitude land was covered by thick ice (Fig. 11.4), with the result that little vegetation was growing there for most of this time.

Fig. 11.5 Wetland forests that covered large areas of the palaeotropics during Pennsylvanian times, and which resulted in thick coal deposits. a. General view from air showing levee-bound rivers dissecting the floodplain. b. View of one of the raised levees (right) with pteridosperms and ferns, the swamp behind dominated by arborescent lycophytes (centre-left), and an area of lake fringed by sphenophytes (left). From Thomas & Cleal (1993).

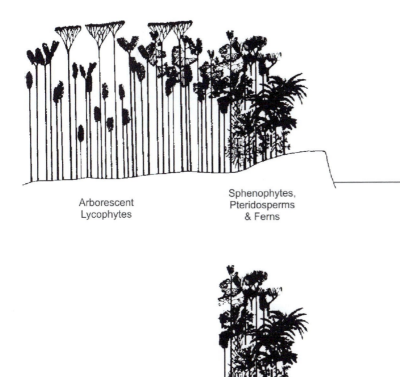

Arborescent
Lycophytes

Sphenophytes,
Pteridosperms
& Ferns

Fig. 11.6 Cross section through a typical levee and adjacent floodbasin area of a coal forest, when the floodbasins were mostly covered by peat-forming forests (upper) and during times of flood (lower). Adapted from Gastaldo (1987).

During temporary interglacials, however, there is evidence of low-diversity vegetation dominated by lycophytes and progymnosperms, rather similar to the vegetation that was found here in Mississippian times.

At one time it was thought that the onset of the glacial conditions was the result of climatic cooling that had been caused by the growth of the palaeotropical forests, through a lowering of atmospheric CO_2. However, it is now clear that the glaciation started well before the forests spread over the palaeotropical lowlands. It seems that the glaciation may have been caused by mountain building in the southern (Gondwana) area, especially in what today is Australia. The snow on these mountains would not have melted in summer so much, resulting in the formation of upland glaciers. These glaciers would have reflected more of the sun's heat back into space, causing a cooling of the climate. This in turn would allow the glaciers to expand,

reflecting more heat back into space, thus producing a positive feedback cycle that would further cool the global climate. Because so much water became locked-up in these glaciers, sea-levels fell to expose areas of continental shelf. It was on the newly exposed areas of continental shelf in the tropics that the Coal Forests developed.

Although the growth of the Coal Forests may not have caused the glaciation, the dynamics of the two did become linked in later Carboniferous time. Towards the end of the Middle Pennsylvanian Epoch, mountains started to rise up in Europe and eastern North America, causing an increase in the sediment that was being washed into the lowland areas where the Coal Forests grew. The arborescent lycophytes which dominated these forests found it difficult to cope with this change and by the end of Middle Pennsylvanian times the western Coal Forests had all but disappeared. Evidence particularly from Australia and South America suggests that this coincided with a marked contraction in the polar ice, and geochemical evidence from marine deposits in lower palaeolatitudes also indicates a significant increase in global temperatures at this time. It seems likely that because the arborescent lycophytes were such fast-growing plants (see Chapter 5) they were very efficient at drawing CO_2 out of the atmosphere. Consequently, when these forests contracted, atmospheric CO_2 increased, and thus so did global temperatures.

These changes did not mean that there was no vegetation in Late Pennsylvanian times in Europe and North America. Palynology and other evidence suggest there were widespread conifers, cycads and other seed-plants here, as well as some ferns, herbaceous sphenophytes and herbaceous lycophytes. There were even some areas where coal-forming peat accumulated, especially in the USA, but this was mostly formed by ferns and some medullosalean pteridosperms, which were not as fast growing as the arborescent lycophytes and therefore did not extract as much CO_2 from the atmosphere. China and neighbouring areas escaped these changes, and the Coal Forests continued to flourish there through into Permian times.

High northern palaeolatitudes were mainly covered by deep ocean, but there is some evidence of sea-ice having formed at that time. The northern middle palaeolatitudes contained relatively impoverished floras dominated by sphenophytes and primitive pteridosperms (the Angara Palaeokingdom, Fig. 11.4). These floras are subdivided into the more typical Angaran macrofloras of Siberia (the Kuznetsk Palaeoarea), and the macrofloras of Kazakhstan which occupied more southerly palaeolatitudes and show a mixture of taxa from Siberia and from the Amerosinian Palaeokingdom.

Permian Period (251–299 Ma)

This period marks the culmination of the first great development of vegetation on land. Several groups of plants, which made their first tentative appearance in the fossil record in the Carboniferous Period, suddenly become significantly more abundant, including conifers, cycads,

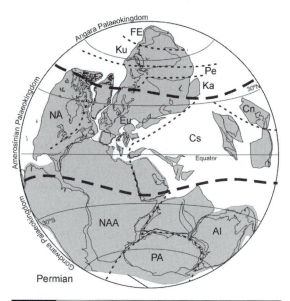

Fig. 11.7 Global distribution of main floristic areas in Permian times. The Amerosinian Palaeokingdom is divided from west to east into North America (NA), Europe (Eu), North Cathaysia (Cn) and South Cathaysia (Cs) palaeoareas. The Angara Palaeokingdom is divided into Kazakhstan (Ka), Pechora (Pe), Kuznetsk (Ku) and Far East (FE) palaeoareas. The Gondwana Palaeokingdom is divided into Nothoafroamerica (NAA), Australoindia (AI) and Palaeoantarctica (PA) palaeoareas. Based on Cleal (1991), using maps by Torsvik et al. (2002).

glossopterids and peltasperms. How much this increase merely reflects the bias in the fossil record towards lowland vegetation is difficult to say. Palynological evidence has shown that conifers, peltasperms and cycads were certainly growing in the areas away from the lowland depositional basins in Pennsylvanian times, but how extensive these forests were is unknown.

Permian times saw the greatest biogeographical diversity among Palaeozoic floras (Fig. 11.7). The low-palaeolatitude Amerosinian Palaeokingdom is subdivided into an east–west series of palaeoareas, probably reflecting physical barriers such as oceans and mountains that hindered plant migrations. In the western tropical areas of Europe and North America, conditions became progressively more arid and conifers, peltasperms and cycads dominated the vegetation. In the east, in China, lycophyte-dominated wetland forests continued to flourish during Early Permian times, but then in Late Permian times underwent a contraction. This led to the progressive decline of many of the characteristic plant groups of these wetland forests, including the arborescent lycophytes and sphenophytes, and the medullosalean pteridosperms.

Glaciation returned to southern Gondwana in Early Permian times, coincident with the expansion of the Coal Forests of China. However, as this glaciation collapsed later in Early Permian times, abundant vegetation spread over the higher palaeolatitudes of both the northern and southern hemispheres. This vegetation resulted in thick deposits of peat, which have produced some of the world's major coal reserves, such as those of Siberia, Kazakhstan, South Africa, India and Australia. The southern middle- and high-palaeolatitude forests of Gondwana were characterised by glossopterid trees, while cordaitanthaleans dominated the northern forests. With this development of middle- and high-latitude forests, we see a marked increase in biogeographical diversity in global vegetation. Although plants could now grow in much higher palaeolatitudes, the conditions of extreme seasonality restricted the number of plant groups that could survive there. In the very high southern latitudes of what is today Antarctica, the forests were dominated almost exclusively by the glossopterids, but in lower palaeolatitudes, in India, South Africa, South America and Australia, there were also sphenophytes, peltasperms, cycads and conifers. Similarly, in the very high northern palaeolatitudes of what is today north-eastern Siberia, the forests were largely dominated by cordaitanthaleans, while the middle-palaeolatitudes, such as in central and western Siberia, Mongolia and Kazakhstan, supported more diverse forests. This more complex floristic structure of the Permian middle and high palaeolatitude vegetation is reflected in the palaeoareas that have been identified in the plant fossil record of this time.

Permian/Triassic (P/T) Extinction Event

The end of the Permian saw the most dramatic disruption ever to occur to the world's vegetation. The tropical swamp forests, already in decline, finally disappeared along with their dominant plants, the arborescent lycophytes. At the same time, the higher latitudes lost the cordaitanthaleans and glossopterids. It was all part of the massive Permian/Triassic boundary extinction event, which caused some 96% of the known species of both animal, and plant, to become extinct. Of the 22 plant families known from the Permian, only nine range through into the Triassic Period (Fig. 11.8). The effects of the extinction event can be seen in the fossil record throughout the world and clearly represented a cataclysmic event that came close to exterminating all life on the planet.

Despite the immensity of the biotic change that occurred at this time, we are still very much in the dark as to its causes. Despite searches for evidence of asteroid or comet impact, such as the iridium layer at the Cretaceous-Tertiary boundary (see below), an extra-terrestrial explanation seems unlikely. One suggestion has been that a sudden change in oceanic circulation at the end of the Permian resulted in stagnant deep-oceanic waters, rich in carbon dioxide and hydrogen sulphide, being brought to near the surface. The resulting surge in these gases into the atmosphere (known as the 'big belch') would have had the combined effect of global climatic warming and depleting oxygen levels in the atmosphere, which together may have been sufficient to have had such a dramatic effect on the Earth's biota. Alternatively, extensive volcanic activity at the end of the Permian Period, which resulted in large areas of flood-basalts in Siberia and pyroclastic deposits in South China, could have emitted noxious gases and enormous quantities of particulates into the atmosphere. These emissions could have altered the atmosphere enough to bring about climate change causing the mass extinction. Whatever its cause, however, this greatest of all mass extinctions resulted in a fundamental change in land vegetation from the primitive Palaeozoic floras to floras of a recognisably modern aspect that appeared in the subsequent Triassic Period.

Triassic Period (200–251 Ma)

The Triassic Period was characterised by relatively high global temperatures, and low precipitation

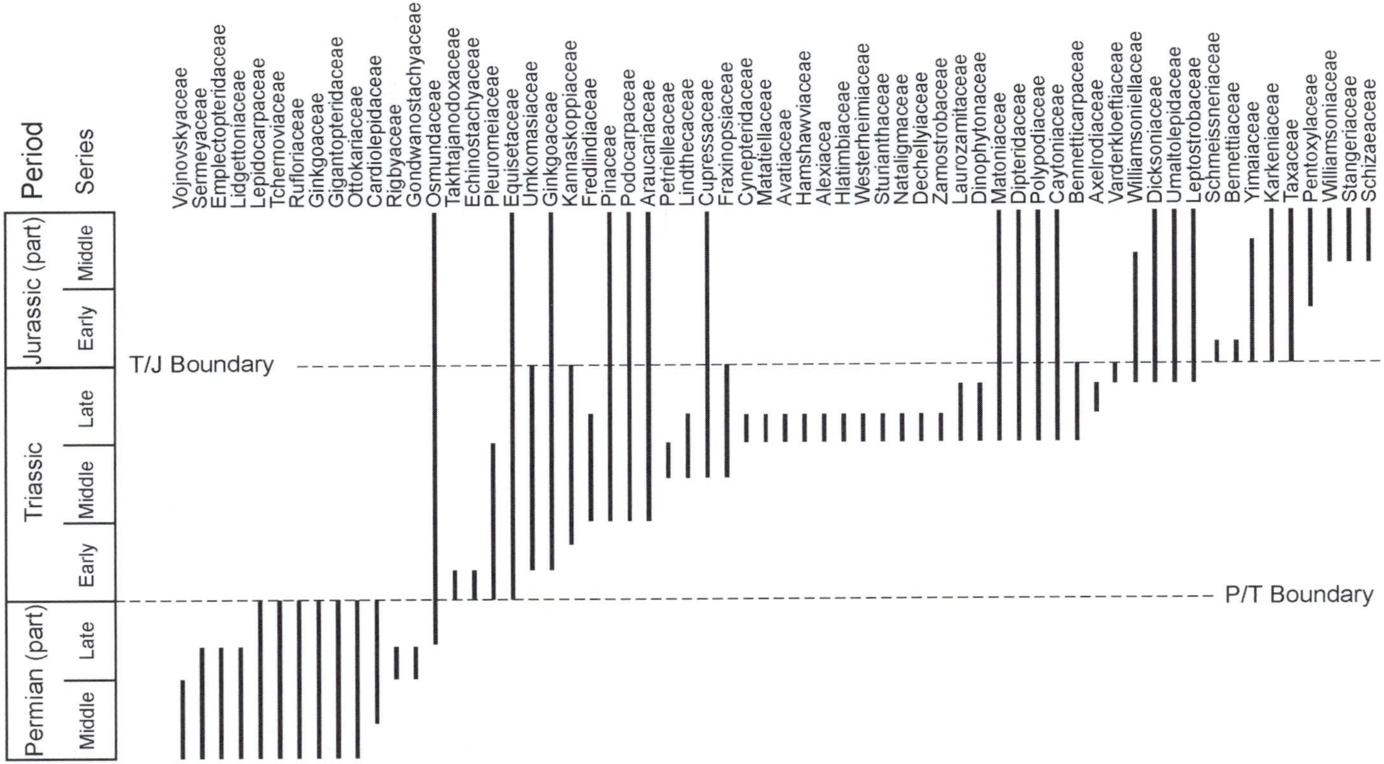

Fig. 11.8 Distribution of plant families between Middle Permian and Middle Jurassic times, showing the effect of the Permian/Triassic (P/T) and Triassic-Jurassic (T/J) boundary events. The Lycopodiaceae, Selaginellaceae, Isoetaceae, Marattiaceae, Gleicheniaceae, Voltziaceae, Peltaspermaceae and Cycadaceae are not shown because they all range through the entire time interval. Ranges based on Benton (1993) and Anderson et al. (2007).

and atmospheric oxygen. This affected the recovery of land vegetation, which remained significantly depleted during Early Triassic times. There is also an almost complete absence of Early Triassic coal deposits throughout the world. The fossil record reveals that the Early Triassic lowland vegetation was dominated by the lycophyte *Pleuromeia* and voltzialean conifers. Conditions started to recover somewhat during Middle Triassic times, and there was a corresponding increase in plant diversity, with the appearance of sphenophytes, cycadophytes, possible bennettitaleans and a range of other enigmatic seed-plants (Fig. 11.7). Importantly, many modern-day families of ferns and conifers started to appear at this time.

It is difficult to identify any biogeographical patterns in Early and Middle Triassic vegetation but by Late Triassic times three clear realms can be recognised. In the tropical palaeolatitudes of Europe, North America and China, diverse floras developed including a range of ferns, sphenophytes, cycads, bennettitaleans, leptostrobaleans, ginkgos and conifers. The general balance of the flora within the palaeotropical belt tended to be relatively uniform with differences, for example between North America and China, being mainly at the species level. Middle palaeolatitudes had generally similar floras, albeit often not as species-rich. Northern middle palaeolatitudes (Angara) were characterised by leptostrobaleans and ginkgoaleans, but bennettitaleans and dipteridacean ferns were absent. Southern middle palaeolatitudes, especially southern Africa and Australia, had diverse floras, particularly notable for the Umkomasiales (*Dicroidium*), but lacking the cheirolepidiacean conifers that occurred elsewhere. On the whole, however, it seems that biogeographical differentiation was relatively weak in the lowland Late Triassic vegetation, probably reflecting a fairly uniform, frost-free climate over much of the world.

Jurassic Period (146–200 Ma)

The end of Triassic times was marked by another mass extinction. What caused it is again uncertain, although some authors have suggested a series of large meteoroid impacts. Its effects on the land fauna were not especially dramatic, but were more so on land vegetation especially the gymnosperms, with 23 of the 48 families found in Triassic floras becoming extinct during the last third of the period (Fig. 11.8). These extinctions were mainly among the bennettitaleans, the groups allied to the peltasperms, and putative gnetaleans, but the conifers were relatively unaffected. This pruning of the gymnosperm diversity effectively laid down the basis of gymnosperm diversity for the rest of Mesozoic times.

Floristic differentiation became more strongly developed than in Triassic times (Fig. 11.9) but still nothing like that seen in the Permian macrofloras. Consequently, Jurassic macrofloras tend to be differentiated as separate palaeoareas rather than palaeokingdoms. As in the Triassic Period, the low palaeolatitudes had the most diverse Early and Middle Jurassic vegetation (western North America, Europe, Central Asia and the Far East – the Equatorial Palaeoarea), which included ferns, sphenophytes, cycads, bennettitaleans, leptostrobaleans, ginkgos and conifers. By now, most ferns belonged to families that are still

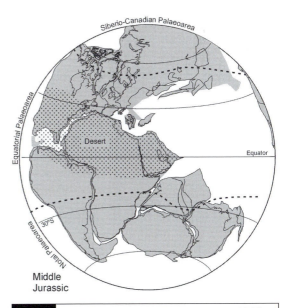

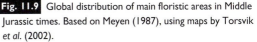
Fig. 11.9 Global distribution of main floristic areas in Middle Jurassic times. Based on Meyen (1987), using maps by Torsvik *et al.* (2002).

living today (e.g. Dipteridaceae, Matoniaceae, Gleicheniaceae, Cyatheaceae). The conifers of this age can also be mostly assigned to modern families (e.g. Podocarpaceae, Araucariaceae, Pinaceae, and Taxaceae), although one of the most abundant families of this time became extinct at the end of the Cretaceous Period (the Cheirolepidiaceae). Coals are often associated with this type of vegetation, reflecting high rainfall, standing water and lush vegetation.

The exception to this pattern was in the western part of the tropical belt (central and eastern North America, and North Africa), where desert conditions prevailed in Early to Middle Jurassic times (Fig. 11.9). Towards the end of the Jurassic Period, this arid area expanded to cover much of the low latitudes. The rare plant fossils in this arid area reflect a largely impoverished vegetation dominated by cheirolepidiacean conifers and matoniacean ferns (*Weichselia*). The desert did not extend as far as northern Europe, but conditions here, nevertheless, became much drier and the dominant plants (bennettitaleans, cycads, peltasperms and cheirolepidiacean conifers) tended to have thick cuticles with sunken stomata, features normally associated with dry conditions. Coals were rarely formed here at this time.

A vegetation-type known as 'Ginkgoalean taiga', mostly dominated by ginkgos and leptostrobaleans, characterised the northern middle palaeolatitudes (the Siberio-Canadian Palaeoarea – Fig. 11.9). Extensive coal deposits were formed here at this time suggesting that conditions were relatively warm and humid. This idea is supported by the virtual absence of cheirolepidiacean conifers that were better adapted to more arid climates. Cycads and bennettitaleans only became relatively common here in Late Jurassic times, notably in eastern Siberia.

The vegetation of the southern palaeolatitudes in Early to Middle Jurassic times (the Notal Palaeoarea) was similar but less diverse than that of the tropical belt, although some plants such as the ginkgos were scarce. In contrast to the northern palaeolatitudes, there were abundant cheirolepidiacean conifers. There was little or no polar ice at this time and so this southern vegetation extended into very high palaeolatitudes, including Antarctica. During Late Jurassic times, the palaeotropical desert expanded into parts of the southern middle palaeolatitudes (South America and Africa) but elsewhere the diversity of vegetation remained high, often including sphenophytes, ferns, cycads, bennettitaleans, ginkgos and conifers. One of the best-known fossil floras of this age is from the Rajmahal Hills in India, which has yielded beautifully preserved petrifactions showing anatomical structure, notably of ferns, cycads, bennettitaleans, ginkgos and conifers.

Cretaceous Period (66–146 Ma)

Early Cretaceous vegetation was broadly similar to that of Late Jurassic times, both in distribution and general composition (Fig. 11.10). Low palaeolatitudes were dominated by arid, desert or sub-desert conditions (South America, central and North Africa, central Asia), where the vegetation was mainly cheirolepidiacean conifers and matoniacean ferns. Northern middle palaeolatitudes, such as in Europe and North America (the European-Sinian Palaeoarea) contained a more

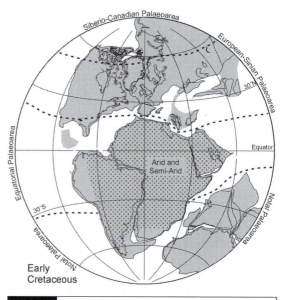

Fig. 11.10 Global distribution of main floristic areas in Early Cretaceous times. Based on Meyen (1987), using maps by Torsvik *et al.* (2002).

Fig. 11.11 View across a Cretaceous river delta in the northern middle palaeolatitudes. Ferns and sphenophytes occupy the foreground, *Tempskya* ferns and ginkgos can be seen left and centre-left, and a bennettite is on the right. From Thomas & Cleal (1998). Painting by A. Townsend.

diverse vegetation including ferns, bennettitaleans, cycads, caytonias, peltasperms, cheirolepidiacean and cupressacean conifers, and some ginkgos, but with only rare leptostrobaleans (Fig. 11.11). Further north (Siberio-Canadian Palaeoarea), the diversity again declined, to be replaced by leptostrobalean- and ginkgo-dominated assemblages. The southern middle palaeolatitude floras (Notal Palaeoarea) were dominated by bennettitaleans and cheirolepidiacean conifers.

Late Cretaceous times saw the start of a major change in style of vegetation, marked mainly by the proliferation of the angiosperms (see Chapter 10) and the corresponding decline in some of the typical Mesozoic elements, such as leptostrobaleans, bennettitaleans, ginkgos and cycads. In effect, it marks the beginning of the end for the typical gymnosperm-dominated Mesozoic flora, and the onset of the development of today's angiosperm-dominated vegetation.

During Late Cretaceous times, arid conditions continued to extend over most of the palaeotropics. What little is known of the palaeotropical vegetation of this time (from northern South America, central Africa and India) shows it to have been dominated by palms and proteas, with few if any conifers. Palms also extended into northern middle palaeolatitudes. The most diverse floras of this time were in North America and were dominated by evergreen angiosperms. There were still some conifers, especially the redwood *Sequoia*, which may have been the dominant plants in some areas.

The absence of polar ice meant that angiosperms extended into high palaeolatitudes at this time, and their remains have been found within a few degrees of the North Pole in Siberia (Fig. 11.10). In contrast to the trees of the middle palaeolatitudes, it is not surprising that these high palaeolatitude angiosperms were deciduous. Conifers were also more common at these higher palaeolatitudes especially the cupressaceans (*Sequoia, Metasequoia*). Relicts of more typical Mesozoic vegetation with caytonias, bennettitaleans and leptostrobaleans were still present in places in the high palaeolatitudes notably Siberia. Elsewhere, however, these essentially Mesozoic groups declined in abundance through the Cretaceous Period.

We know relatively little about the Late Cretaceous vegetation of the southern middle palaeolatitudes, which were largely covered by the same desert that affected the low palaeolatitudes. What evidence is available suggests that sphenophytes, ferns, conifers and angiosperms dominated the vegetation fringing the deserts. The most abundant conifers were araucariaceans and podocarps, while the angiosperms included *Nothofagus* (southern beech), all of which are typical southern hemisphere plants of today. As in the northern hemisphere, conifer and broad-leaved angiosperm forests seem to have extended into very high southern palaeolatitudes, suggesting there was little if any polar ice.

Cretaceous/Tertiary (K/T) Extinction Event

In several places in the world, the boundary between the Cretaceous and Palaeogene Systems (widely referred to as the K/T boundary) is marked by a very distinctive layer of clay with unusually high levels of iridium (Fig. 11.12). This high level of iridium, which is normally a very rare element on the surface of the Earth, is thought to be the result of either a boloid (asteroid) impact, probably near Mexico, or massive volcanic activity in India (or possibly both). Volcanic activity can certainly produce enormous quantities of ash that can circle the earth for weeks before settling. The largest explosive eruption in historic times was of Tambora, Indonesia in 1815 which launched about 43 000 million tons of ash into the sky, while it is estimated that when Mount Manza in Oregon exploded 5000 years ago it released about 54 000 million tons. These were minor explosions compared to the ones we suspect happened in India. Whatever the cause, it seems that the end of the Cretaceous Period was marked by a period of darkness and lowered temperatures.

It is now well-known that the dinosaurs became extinct at this time, as did some marine animals such as the ammonites and some planktonic foraminifera. However, the event had less effect on plant life. Very few plant-groups became extinct at this time, and those that did (e.g. the bennettitaleans and caytonias) were already in significant decline during the Cretaceous Period, probably as a result of being out-competed by the angiosperms. The darkness and lower temperatures would undoubtedly have affected plants, killing both their vegetative and reproductive organs. Those flowering plants that had never been exposed to low temperatures would have been particularly susceptible.

Fig. 11.12 The K/T boundary layer exposed near Raton, New Mexico, USA. The white layer near the hammer is the iridium-rich bed that marks the junction between the Mesozoic and the Cenozoic. Photograph by R. A. Spicer (Milton Keynes). From Thomas & Cleal (1998).

However, the seeds and spores that had already been released from the plants, and which were lying dormant on or in the ground, would mostly have been unaffected by the environmental disruption and would have been capable of germination when conditions improved.

The temporary effect of this sudden 'ecological shock' can most readily be seen in the fossil record from the middle palaeolatitudes of present-day western North America. Immediately above the K/T boundary there is a marked dominance of ferns in the pollen and spore record. This is analogous to the early colonisation of land laid bare by volcanic activity such as in Krakatau (Indonesia) in 1883, Mount St Helens (USA) in 1980 and Mount El Chichon (Mexico) in 1982. The subsequent Palaeogene successional colonisation in North America showed a selection in favour of deciduous taxa, and an elimination of the evergreen vegetation that was dominant before the trauma. This catastrophe has been interpreted as a short freeze of possibly less than a year, rather than a longer impact winter, because it did not affect the warmer and more southerly floras, or the more northerly ones that were already adapted for deciduousness. The effect certainly appears to have been relatively local because no comparable vegetation trauma has been documented elsewhere. Because there is no evidence for a mass kill other than in the middle latitudes in North America, it was certainly not a global event.

Palaeogene and Neogene Periods (1.8–67 Ma)

These two periods used to be combined and referred to as the Tertiary Period. As there are close similarities between Palaeogene and Neogene vegetation, we will refer to them here as Tertiary vegetation. The first vegetation types characteristic of the modern world appeared in Tertiary times. Many Tertiary plant fossils are identified closely with living taxa and often included within their systematic groupings such that extant genera are commonplace. The fossil record of Tertiary plants has also become linked to the concept of ecological comparisons

and the migration or extinction of plants due to climatic change. This is, of course, too simplistic unless the climatic changes are rapid, because plants can also respond through phylogenetic change.

Latitudinal differences in floras was apparent in early Tertiary times. In North America, the mass-kill of the vegetation at the K/T boundary was followed by a period of increased rainfall, making the conditions right for the widespread development of multi-stratal rainforests in the more southerly areas. This development is identified in the fossil leaf floras by an increase in leaf size, the high percentage of species with apical drip-tips (a feature of many of today's tropical rainforest plants) and vines. Northwards there were temperate broad-leaved evergreen forests and mixed conifer forests. Finally there was a broad-leaved deciduous forest, the polar Arcto-Tertiary forest, found on the north-west Canadian archipelago at a palaeolatitude of 75°–80° north, where mild moist summers at about 25°C alternated with continuous winter darkness at about 0°–4°C. This primary deciduous vegetation included members of the Platanaceae, Juglandaceae, Betulaceae, Menispermaceae, Cercidophyllaceae, Ulmaceae, Fagaceae and Magnoliaceae with gymnosperms belonging to the Cupressaceae, Pinaceae and the Ginkgoaceae. This flora was able to spread between Europe and North America until the land bridge between them was severed in the Palaeocene. Similar interchange between North America and Asia was possible by the land bridge in the Bering Strait between late Cretaceous and late Neogene times. In North America a mid-continental seaway separated the European integration from that with Asia.

A different, more southern flora was growing around the Tethys Sea, the southern USA and Asia. This lower latitude flora has been described as being of a tropical type because many plants have close affinities with those growing in today's tropics. Much of the information comes from a study of seed and fruit floras found in Europe and North America, notably the London Clay flora which includes about 350 species. The interpretation of the vegetation is based on an assessment of the nearest

living relatives and their ecological tolerances. Although some of the ecological tolerances may have changed over the last 50 million years, this approach does provide some indication of the palaeoclimate. Many of the London Clay species have their nearest living relatives among the tropical vegetation of the Indo-Asian region but there are many others whose affinities are with living temperate species. The London Clay flora, like other Tertiary floras, cannot be precisely equated to any flora or climatic condition that exists today. The closest analogue is a semi-tropical, or pantropical, east Asian rainforest which contained all the elements of the tropical rainforest except the taller dipterocarp trees that stand above the main forest canopy. The dense forest had abundant shrubs and climbers with the plants of 'temperate affinity' growing along the stream sides and in open habitats within the forest. The shoreline and the banks of the larger rivers would have been covered with mangrove which was dominated over large areas by the stemless-palm *Nypa*. Reed-swamp with plants similar to living *Scirpus* bordered the rivers above the reach of tidal influence. Today tropical forests delimited by the 20° and 25°C mean annual temperature isotherms are found in South America, Africa and Asia (Fig. 11.13).

Within two million years, many of these 'tropical' elements had disappeared from Europe, having migrated southwards. At this time, the Arcto-Tertiary flora was fragmenting and it migrated southwards during the Eocene/Oligocene cooling and the onset of the Pleistocene fluctuations. For the first time in the history of the angiosperms, there were large areas of temperate deciduous forests. In Europe, waves of Arcto-Tertiary elements such as *Liquidamber*, *Ulmus*, *Alnus*, *Populus* and *Acer* penetrated the primary evergreen forests that were defined by laurophyllous elements. This was more pronounced during middle and late Eocene times, when there were mixed forests with much higher numbers of deciduous species. The best surviving examples of the European earlier Tertiary floras are now confined to the wetter parts of the Canary Islands. In North America, the Arcto-Tertiary floras remained to the east of the mid-continental seaway and even now, the floras of the south east USA have more common elements with Europe than with those of western North America. Remnants of the much more extensive *Taxodium* swamps now only persist in southeastern USA (Fig. 11.14).

During Miocene times, there was a marked change in the vegetation in many areas. Open habitats increased in which grasses and other herbaceous plants flourished. This was the first time that prairie vegetation developed and it enabled herbivorous mammals to form large herds that roamed freely across them.

Fig. 11.14 *Taxodium* swamp in Florida, USA. Photo by B. A. Thomas.

Quaternary Period (1.8 Ma to present)

There were dramatic changes in climate during the Quaternary Period. The climate alternated between times of extreme cold, called glacials or ice ages, when much of north west Europe, Siberia and northern North America was covered with ice, and times of relative warmth, called interglacials, when the climate was similar to, or sometimes warmer than, today (Fig. 11.15). There are fossil wood, leaves and fruits found in Quaternary deposits, but our knowledge of Quaternary floras mostly comes from studies of pollen grains and spores recovered from peat, soil, or lake sediments. Samples prepared from cores taken from these sediments show the

relative abundance of different types of pollen and spores which indicates the type of vegetation existing at the time. Pollen assemblages taken from different depths in a core can be compared to reveal how the vegetation in that area changed over time. A comparison of cores from a widespread area shows the vegetational changes over the whole region. These studies reveal that the Quaternary climatic oscillations resulted in large-scale changes in distribution of plant communities, and migrations of animal species southwards and northwards, all in response to cooling and glacial advance and warming and glacial retreat. Many plants became extinct in Europe because deteriorating climatic conditions

in the north gradually restricted the areas where they could grow and reproduce. They could not spread southward because the routes were blocked by the ice covered mountain chains of the Pyrenees, the Alps, the Carpathians and the Urals. In North America this was not the case because the main mountain ranges run north/south thereby being no barrier to north/south migrations.

The final migration of plant species after the last ice age, which reached its maximum extent 18–20 000 years ago, led to the present geographical distribution of new vegetational assemblages throughout the world. Many northern areas now have significantly fewer species

than once lived there, especially where new seas surround isolated islands, such as the British Isles, that were once part of a continental land mass. The migration also led to some unusual distributions of species. There are arctic-alpine species that have disjunct arctic and continental mountain top distributions (Fig. 11.16). This separation came about with climatic amelioration when some plants migrated northwards while others moved higher up the mountains where they eventually became marooned. Mixing plants from originally different habitats like this can lead to interbreeding, perhaps the best documented example of this being in the fern flora in Japan. Species brought south through climatic change then moved up the mountains as the climate ameliorated and freely hybridised with those already living on the mountains. Such hybridisation must have happened many times in the past as climates changed and species migrated and could have been the driving force behind many evolutionary changes.

The study of post-glacial fossil pollen and spores allows us to follow the course of vegetational change brought about through human interference. The most favourable climate for plant distribution was about 7000 years ago when tree cover was at its maximum extent spreading much further northwards and higher into the mountains than is possible today. Climatic deterioration led to some retreat, but by then the selective woodland clearance for agriculture was starting to take effect. Such clearance continued around much of the world with the obvious effects that we can see today. That is, however, outside the scope of this book.

Recommended reading

Anderson *et al.* (2007), Behrensmeyer *et al.* (1992), Benton (1993), Birks & Birks (1980), Boulter & Fischer (1994), Cleal & Thomas (1995), Cleal *et al.* (2001), Dobruskina (1994), Edwards & Wellman (2001), Erwin (1993), Gensel & Andrews (1984), Long (1996), Meyen (1987), Seward (1914), Vakhrameev *et al.* (1978), Wolfe (1997).

References

Abbott, M. L. 1958. The American species of *Asterophyllites, Annularia*, and *Sphenophyllum*. *Bulletin of American Paleontology*, **38**, 289–390.

Anderson, J. M. & Anderson, H. M. 2003. *Heyday of the Gymnosperms: Systematics and Biodiversity of the Late Triassic Molteno Fructifications*. Sterlitzia, Volume 15. Pretoria SA: National Botanic Institute.

Anderson, J. M., Anderson, H. M. & Cleal, C. J. 2007. *Brief History of the Gymnosperms: Classification, Biodiversity, Phytogeography and Ecology*. Sterlitzia, Volume 20. Pretoria SA: National Botanic Institute.

Andrews, H. N. 1952. A fossil osmundaceous tree fern from Brazil. *Bulletin of the Torey Botanical Club*, **77**, 29–34.

Andrews, H. N. 1980. *The Fossil Hunters: In Search of Ancient Plants*. Ithaca, NY: Cornell University Press.

Arber, E. A. N. 1921a. A sketch of the history of palaeobotany with special reference to the fossil flora of the British Coal Measures. In C. Singer, ed., *Studies in the History and Method of Science*. Oxford: Clarendon Press, pp. 471–489.

Arber, E. A. N. 1921b. *Devonian Floras: A Study of the Origin of Cormophyta*. Cambridge: Cambridge University Press.

Ash, S. R. & Pigg, K. B. 1991. A new Jurassic *Isoetites* (Isoetales) from the Wallow Terrane in Hell's Canyon, Oregon and Idaho. *American Journal of Botany*, **78**, 1636–1642.

Axsmith, B. J., Krings, M. & Taylor, T. N. 2001. A filmy fern from the Upper Triassic of North Carolina (USA). *American Journal of Botany*, **88**, 1558–1567.

Banks, H. P. 1964. *Evolution and Plants of the Past*. London: Macmillan.

Bashforth, A. R. & Zodrow, E. L. 2007. Partial reconstruction and palaeoecology of *Sphenophyllum costae* (Middle Pennsylvanian, Nova Scotia, Canada). *Bulletin of Geosciences*, **82**, 365–382.

Bateman, R. M. 1991. Palaeobiological and phylogenetic implications of anatomically preserved *Archaeocalamites* from the Dinantian of Oxroad Bay and Loch Humphrey Burn, southern Scotland. *Palaeontographica, Abteilung B*, **223**, 1–59.

Bateman, R. M., DiMichele, W. D. & Willard, D. A. 1992. Experimental cladistic analysis of anatomically preserved arborescent lycopsids from the Carboniferous of Euamerica: an essay on paleobotanical phylogenetics. *Annals of the Missouri Botanical Garden*, **79**, 500–559.

Batenburg, L. H. 1977. The *Sphenophyllum* species in the Carboniferous flora of Holz (Westphalian D, Saar Basin, Germany). *Review of Palaeobotany and Palynology*, **24**, 69–99.

Batenburg, L. H. 1981. Vegetative anatomy and ecology of *Sphenophyllum zwickaviense, S. emarginatum*, and other 'compression species' of *Sphenophyllum*. *Review of Palaeobotany and Palynology*, **32**, 275–313.

Batten, D. J., Collinson, M. E. & Knobloch, E. 1994. *Ariadaesporites* and *Capulisporites*: 'water fern' megaspores from the Upper Cretaceous of Central Europe. *Review of Palaeobotany and Palynology*, **83**, 159–174.

Beck, C. B. 1960. The identity of *Archaeopteris* and *Callixylon*. *Brittonia*, **12**, 351–368.

Beck, C. B. (ed.) 1976. *Origin and Early Evolution of Angiosperms*. New York: Columbia University Press.

Beck, C. B. (ed.) 1988. *Origin and Evolution of Gymnosperms*. New York: Columbia University Press.

Behrensmeyer, A. K., Damuth, J. D., DiMichele, W. A., et al. (eds.) 1992. *Terrestrial Ecosystems Through Time*. Chicago, IL: University of Chicago Press.

Benton, M. J. (ed.) 1993. *The Fossil Record 2*. London: Chapman & Hall.

Birks, H. J. B. & Birks, H. H. 1980. *Quaternary Palaeoecology*. London: Edward Arnold.

Boersma, M. & Broekmeyer, L. M. 1979. *Index of Figured Plant Megafossils. Carboniferous 1971–1975*. Laboratory of Palaeobotany and Palynology, University of Utrecht (Special Publication No. 1).

Boersma, M. & Broekmeyer, L. M. 1980. *Index of Figured Plant Megafossils. Triassic 1971–1975*. Laboratory of Palaeobotany and Palynology, University of Utrecht (Special Publication No. 2).

Boersma, M. & Broekmeyer, L. M. 1981. *Index of Figured Plant Megafossils. Permian 1971–1975*. Laboratory of Palaeobotany and Palynology, University of Utrecht (Special Publication No. 3).

Boersma, M. & Broekmeyer, L. M. 1982. *Index of Figured Plant Megafossils. Jurassic 1971–1975*. Laboratory of Palaeobotany and Palynology, University of Utrecht (Special Publication No. 4).

Boulter, M. C. & Fisher, H. C. (eds.) 1994. *Cenozoic Plants and Climates of the Arctic*. Heidelberg: Springer Verlag.

Bowden, A. J., Burek, C. V. & Wilding, R. (eds.) 2005. History of palaeobotany: selected essays. *Geological Society of London, Special Publication*, **241**, 1–304.

Boureau, E. 1964. *Traité de Paléobotanique, 3. Spheno-phyta, Noeggerathophyta.* Paris: Masson et Cie.

Boureau, E. 1971. *Les Sphénophytes. Biologie et Histoire Évolutive.* Paris: Vuibert.

Bowerbank, J. S. 1840. *A History of the Fossil Fruits and Seeds of the London Clay.* London: John Van Voorst.

Brack-Hanes, S. D. & Greco, A. M. 1988. Biomineralization in *Thalassia testudinum* (Liliopsida: Hydrocharitaceae) and an Eocene seagrass. *Transactions of the American Microscopical Society*, **107**, 286–292.

Brack-Hanes, S. D. & Thomas, B. A. 1983. A re-examination of *Lepidostrobus* Brongniart. *Botanical Journal of the Linnean Society*, **86**, 125–133.

Brongniart, A. 1828. Prodrome d'une histoire des végétaux fossiles. *Dictionnaire des Sciences Naturelles*, **57**, 1–223.

Brongniart, A. 1828–1838. *Histoire des Végétaux Fossils*, 15 parts. Paris & Strasbourg: F. G. Levrault.

Brousmiche, C. 1983. Les fougères sphénoptéridiennes du bassin houiller Sarro-Lorraine. *Publication Société Géologique du Nord*, **10**, 1–480.

Brown, R. W. 1956. Palm-like plants from the Dolores formation (Triassic) South-western Colorado. *US Geological Survey Professional Paper*, **274-H**, 205–209.

Brunton, C. H. C., Besterman, T. P. & Cooper, J. A. 1985. Guidelines for the curation of geological materials *Geological Society of London, Miscellaneous Paper* **17**.

Burek, C. V. & Higgs, B. (eds.) 2007. The role of women in the history of geology. *Geological Society of London, Special Publication*, **281**, 1–342.

Burek, C. V. & Prosser, C. D. (eds.) 2008. The history of geoconservation. *Geological Society of London, Special Publication*, **300**, 1–312.

Burgh, J. van der 1993. Oaks related to *Quercus petraea* from the Upper Tertiary of the Lower Rhenish Basin. *Palaeontographica, Abteilung B*, **230**, 195–201.

Call, V. B. & Dilcher, D. L. 1992. Investigations of angiosperms from the Eocene of southeastern North America: samaras of *Fraxinus wilcoxiana* Berry. *Review of Palaeobotany and Palynology*, **74**, 249–266.

Camus, J. M. (ed.) 1991. The history of British pteridology 1891–1991. *The British Pteridological Society Special Publication*, **4**, 7–15.

Camus, J. M., Gibby, M. & Johns, R. J. (eds.) 1996. *Pteridology in Perspective.* London: Royal Botanic Gardens, Kew.

Camus, J. M., Jermy, A. C. & Thomas, B. A. 1991. *A World of Ferns.* London: Natural History Museum.

Chaloner, W. G. & Macdonald, P. 1980. *Plants Invade the Land.* Edinburgh: HMSO.

Chandra, S. & Surange, K. R. 1979. Revision of the Indian species of *Glossopteris. Birbal Sahni Monograph*, **2**, 1–291.

Cleal, C. J. (ed.) 1991. *Plant Fossils in Geological Investigation: The Palaeozoic.* Chichester, UK: Ellis Horwood.

Cleal, C. J. 2005. The Westphalian macrofloral record from the cratonic central Pennines Basin, UK. *Zeitschrift der Deutschen Gesellschaft für Geowissenschaften*, **156**, 387–410.

Cleal, C. J. 2008. Palaeofloristics of Middle Pennsylvanian medullosaleans in Variscan Euramerica. *Palaeogeography, Palaeoecology, Palaeoclimatology*, **268**, 164–180.

Cleal, C. J. & Shute, C. H. 1995. A synopsis of neuropteroid foliage from the Carboniferous and Lower Permian of Europe. *Bulletin of the British Museum (Natural History), Geology Series*, **51**, 1–52.

Cleal, C. J. & Thomas, B. A. 1994. *Plant Fossils of the British Coal Measures, Field Guide to Fossils 6.* London: Palaeontological Association.

Cleal, C. J. & Thomas, B. A. 1995. *Palaeozoic Palaeobotany of Great Britain, Geological Conservation Review Series, No. 9.* London: Chapman & Hall.

Cleal, C. J. & Thomas, B. A. 1999. *Plant Fossils, Fossils Illustrated, Volume 2.* Woodbridge, UK: Boydell Press.

Cleal, C. J., Thomas, B. A., Batten, D. J. & Collinson, M. E. 2001. *Mesozoic and Tertiary Palaeobotany of Great Britain, Geological Conservation Review Series, No. 22.* Peterborough, UK: Joint Nature Conservation Committee.

Collinson, M. E. 1980. A new multiple-floated *Azolla* from the Eocene of Britain with a brief review of the genus. *Palaeontology*, **23**, 213–229.

Collinson, M. E. 1983. *Fossil Plants of the London Clay, Field Guide to Fossils, No. 1.* London: Palaeontological Association.

Collinson, M. E. 2001. Cainozoic ferns and their distribution. *Brittonia*, **53**, 173–235.

Collinson, M. E. 2002. The ecology of Cainozoic ferns. *Review of Palaeobotany and Palynology*, **119**, 51–68.

Collinson, M. E., Kvaček, Z. & Zastawniak, E. 2001. The aquatic plants *Salvinia* (Salviniales) and *Limnobiophyllum* (Arales) from the Late Miocene flora of Sośnica (Poland). *Acta Palaeobotanica*, **41**, 253–282.

Cornish, L. & Doyle, A. 1984. Use of ethamolamine thioglycollate in the conservation of pyritized fossils. *Palaeontology*, **27**, 421–424.

Crabb, P. 2001. The use of polarised light in photography of macrofossils. *Palaeontology*, **44**, 659–664.

Crane, P. R. 1982. Betulaceous leaves and fruits from the British Upper Palaeocene. *Botanical Journal of the Linnean Society*, **83**, 103–136.

Crane, P. R. 1989. Paleobotanical evidence on the early radiation of nonmagnoliid dicotyledons. *Plant Systematics and Evolution*, **162**, 165–191.

Crane, P. R. & Blackmore, S. 1989. *Evolution, Systematics, and Fossil History of the Hamamelidae*, Special Volume 40, 2 vols. London: Systematics Association.

Crane, P. R. & Dilcher, D. L. 1985. *Lesqueria*: an early angiosperm fruiting axis from the mid-Cretaceous. *Annals of the Missouri Botanical Garden*, **71**, 384–402.

Crawley, M. 1989. Dicotyledonous wood from the Lower Tertiary of Britain. *Palaeontology*, **32**, 597–622.

Crepet, W. L., Friis, E. M. & Nixon, K. C. 1991. Fossil evidence for the evolution of biotic pollination. *Philosophical Transactions of the Royal Society of London B*, **333**, 187–195.

Cridland, A. A. & Williams, J. L. 1966. Plastic and epoxy transfers of fossil plant compressions. *Bulletin of the Torrey Botanical Club*, **93**, 311–322.

Darwin, C. R. 1839. *Journal of Researches into the Geology and Natural History of the Various Countries During the Voyage of H.M.S. Beagle Round the World*. London: Ward Lock.

Darwin, C. R. 1844. *Geological Observations on the Volcanic Islands and Parts of South America Visited During the Voyage of H.M.S. Beagle*. London: Smith, Elder & Co.

Delevoryas, T. 1966. *Morphology and Evolution of Fossil Plants*. Austin, TX: Holt, Reinhart & Winston.

Dilcher, D. L. 1973. A paleoclimatic interpretation of the Eocene floras of southeastern North America. In A. Graham, ed., *Vegetation and Vegetational History of Latin America*. Amsterdam: Elsevier.

DiMichele, W. A. & Phillips, T. L. 1985. Arborescent lycopod reproduction and paleoecology in a coal-swamp environment of late Middle Pennsylvanian Age (Herrin Coal, Illinois, U.S.A.). *Review of Palaeobotany and Palynology*, **44**, 1–26.

DiMichele, W. A., Phillips, T. L. & Pfefferkorn, H. W. 2006. Paleoecology of Late Paleozoic pteridosperms from tropical Euramerica. *Journal of the Torrey Botanical Society*, **133**, 83–118.

DiMichele, W. A. & Skog, J. E. (eds.) 1992. The Lycopsida: a symposium. *Annals of the Missouri Botanical Garden*, **79**, 447–736.

Dix, E. 1934. The sequence of floras in the Upper Carboniferous, with special reference to South Wales. *Transactions of the Royal Society of Edinburgh*, **57**, 789–838.

Dobruskina, I. A. 1994. Triassic floras of Eurasia. *Österrieiche Akademie der Wissenschaften, Schriftenreihe der Erdwissenschaften Kommissionen*, **10**.

Doyle, P., Bennett, M. R. & Baxter, A. N. 2001. *The Key to Earth History: An Introduction to Stratigraphy*, 2nd edn. London: J. Wiley.

Doyle, J. A. & Hickey, L. J. 1976. Pollen and leaves from the Mid Cretaceous Potomac Group and their bearing on early angiosperm evolution. In C. B. Beck, ed., *Origins and Early Evolution of Angiosperms*. New York: Columbia University Press, pp. 139–206.

Doyle, P. & Robinson, E. 1993. The Victorian 'Geological Illustrations' of Crystal Palace Park. *Proceedings of the Geologists' Association*, **104**, 181–194.

Dyer, A. F. & Page, C. N. (eds.) 1985. Biology of pteridophytes. *Proceedings of the Royal Society of Edinburgh, Section B*, **86**, 1–474.

Edwards, D. 1994. Towards an understanding of pattern and process in the growth of early vascular plants. In D. S. Ingram & A. Hudson, eds., *Shape and Form in Plants and Fungi*. Linnean Society Symposium Series, **16**, 39–59.

Edwards, D. 1996. New insights into early land ecosystems: a glimpse of a Lilliputian world. *Review of Palaeobotany and Palynology*, **90**, 159–174.

Edwards, D. 1997. Charting diversity in early land plants: some challenges for the next millennium. In K. Iwatsuki & P. H. Raven, eds., *Evolution and Diversification of Land Plants*. Tokyo: Springer, pp. 3–26.

Edwards, D., Davies, K. L. & Axe, L. 1992. A vascular conducting strand in the early land plant *Cooksonia*. *Nature*, **357**, 683–685.

Edwards, D. & Wellman, C. H. 2001. Embryophytes on land: the Ordovician to Lochkovian (Lower Devonian) record. In P. G. Gensel & D. Edwards, eds., *Plants Invade the Land: Evolutionary and Environmental Perspectives*. New York: Columbia University Press, pp. 3–28.

Elkund, H., Friss, E. M. & Pedersen, K. R. 1997. Chloranthaceous floral structures from the Late Cretaceous of Sweden. *Plant Systematics and Evolution*, **207**, 13–42.

Eriksson, O., Friss, E. M. & Löfgren, P. 2000. Seed size, fruit size and dispersal systems in angiosperms from the Early Cretaceous to the Late Tertiary. *The American Naturalist*, **156**, 47–58.

Erwin, D. H. 1993. *The Great Paleozoic Crisis: Life and Death in the Permian*. New York: Columbia University Press.

Erwin, D. M. & Stockey, R. A. 1989. Permineralised monocotyledons from the Middle Eocene Princeton chert (Allenby Formation) of British Columbia: Alismataceae. *Canadian Journal of Botany*, **67**, 2636–2645.

Erwin, D. M. & Stockey, R. A. 1991. Silicified mono-cotyledons from the Middle Eocene Princeton Chert (Allenby Formation) of British Columbia, Canada. *Review of Palaeobotany and Palynology*, **70**, 147–162.

Esau, K. 1960. *Plant Anatomy*. London: John Wiley & Sons Ltd.

Florin, R. 1951. Evolution in cordaites and conifers. *Acta Horti Bergiana*, **15**, 285–388.

Florin, R. 1963. The distribution of conifers and taxad genera in time and space. *Acta Horti Bergiana*, **20**, 121–312.

Forey, P. L., Humphries, C. J., Kitching, I. J., *et al.* 1998. *Cladistics: A Practical Course in Systematics*, Vol 10. London: The Systematics Association.

Francis, J. E. 1984. The seasonal environment of the Purbeck (Upper Jurassic) fossil forests. *Palaeogeography, Palaeoclimatology, Palaeoecology*, **48**, 285–307.

Friis, E. M. 1985. Structure and function in Late Cretaceous flowers. *Det Kongelige Danske Videnskabernes Selskab Biologiske Skrifter*, **25**, 1–37.

Friis, E. M. 1985. Angiosperm fruits and seeds from the Middle Miocene of Jutland (Denmark). *Kongelige Danske Videnkaberne Selskab Biologiske Skrifter*, **24**: **3**, 1–65.

Friis, E. M., Chaloner, W. G. & Crane, P. R. (eds.) 1987. *The Origins of Angiosperms and their Biological Consequences*. Cambridge: Cambridge University Press.

Friis, E. M., Crane, P. R. & Pedersen, K. R. 1997. Fossil history of magnoliid angiosperms. In K. Iwatsuki & P. H. Raven, eds., *Evolution and Diversification of Land Plants*. Tokyo: Springer, pp. 121–156.

Friis, E. M. & Endress, P. K. 1990. Origin and evolution of angiosperm flowers. *Advances in Botanical Research*, **17**, 99–162.

Friis, E. M., Pedersen, K. R. & Crane, P. R. 1994. Angiosperm floral structures from the Early Cretaceous of Portugal. *Plant Systematics and Evolution*, **8** (Suppl.), 31–49.

Friis, E. M. & Skarby, A. 1982. *Scandianthus* gen. nov., angiosperm flowers of saxifragalean affinity from the Upper Cretaceous of southern Sweden. *Annals of Botany*, **50**, 569–583.

Gao, Z. & Thomas, B. A. 1989. A review of fossil cycad evidence of *Crossozamia* Pomel and its associated leaves from the Lower Permian of Taiyuan, China. *Review of Palaeobotany and Palynology*, **60**, 205–223.

Gastaldo, R. A. 1981. Taxonomic considerations for Carboniferous coalified compression equisetalean strobili. *American Journal of Botany*, **68**, 1319–1324.

Gastaldo, R. A. 1986. An explanation for lycopod con-figuration, 'Fossil Grove' Victoria Park, Glasgow. *Scottish Journal of Geology*, **22**, 77–83.

Gastaldo, R. A. 1987. Confirmation of Carboniferous clastic swamp communities. *Nature*, **326**, 869–871.

Gastaldo, R. A. 1992. Regenerative growth in fossil horsetails following burial by alluvium. *Historical Biology*, **6**, 203–219.

Gensel, P. G. & Andrews, H. N. 1984. *Plant Life in the Devonian*. New York: Praeger.

Good, C. W. 1971. The ontogeny of Carboniferous artic-ulates: calamite leaves and twigs. *Palaeontographica, Abteilung B*, **113**, 137–158.

Good, C. W. 1975. Pennsylvanian-age calamitean cones, elater-bearing spores, and associated vegetative organs. *Palaeontographica, Abteilung B*, **153**, 28–99.

Gordon, W. T. 1935. Plant life and the philosophy of geology. *Report of the British Association for the Advancement of Science*, **1934**, 49–82.

Hammer, Ø. & Harper, D. 2006. *Paleontological Data Analysis*. Oxford: Blackwell.

Harris, T. M. 1931. The fossil flora of Scoresby Sound East Greenland Part 1: Cryptogram (exclusive of Lycopodiales). *Meddelelser om Grønland*, **85** (2), 1–104.

Harris, T. M. 1932. Fossil flora of Scorseby Sound. *Meddelelser om Grønland*, **85**, 1–112.

Harris, T. M. 1946. Notes on the Jurassic Flora of Yorkshire 31–33. *Annals and Magazine of Natural History, Series 11*, **13**, 392–411.

Harris, T. M. 1961a. The fossil cycads. *Palaeontology*, **4**, 313–323.

Harris, T. M. 1961b–1979. *The Yorkshire Jurassic Flora*, in 5 parts. London: British Museum (Natural History).

Harris, T. M. 1963. Hugh Hamshaw Thomas 1885–1962. *Biographical Memoirs of Fellows of the Royal Society*, **9**, 287–297.

Harris, T. M. 1976. The Mesozoic gymnosperms. *Review of Palaeobotany and Palynology*, **21**, 119–134.

Harris, T. M., Millington, W. & Miller, J. 1974. *The Yorkshire Jurassic Flora. IV. Ginkgoales, Czekanowskiales*. London: British Museum (Natural History).

Hemsley, A. R. 1990. *Parka decipiens* and land plant spore evolution. *Historical Biology*, **4**, 39–50.

Herendeen, P. S, & Dilcher, D. L. (eds.) 1992. *Advances in Legume Systematics Part 4. The Fossil Record*. London: The Royal Botanic Gardens, Kew, 326 pp.

Herendeen, P. S., Less, D. H. & Dilcher, D. L. 1990. Fossil *Ceratophyllum* (Ceratophyllaceae) from the Tertiary of North America. *American Journal of Botany*, **77**, 7–16.

Hickey, L. J. 1973. Classification of the architecture of dicotyledonous leaves. *American Journal of Botany*, **60**, 17–33.

Hill, C. R. 1996. A plant with flower-like organs from the Wealden of the Weald (Lower

Cretaceous), southern England. *Cretaceous Research*, **17**, 27–38.

Hoadley, R. B. 1980. *Understanding Wood: A Craftsman's Guide to Wood*. London: Bell & Hyman in association with Taunton Press.

Hopping, C. A. 1956. A note on the leaf cushions of a species of Palaeozoic arborescent lycopod (*Sublepidophloios venticosus* sp. nov.). *Proceedings of the Royal Society of Edinburgh, Series B*, **66**, 1–9.

Hori, T., Ridge, R. W., Tulecke, W., *et al.* (eds.) 1997. *Ginkgo biloba – a global treasure*. Tokyo: Springer.

Hueber, F. M. 1968. *Psilophyton*: the genus and the concept. In D. H. Oswald, ed., *Symposium on the Devonian System, Volume 2*. Calgary, Canada: Alberta Society of Petroleum Geologists, pp. 815–822.

Hueber, F. M. 1992. Thoughts on the early lycopsids and zosterophylls. *Annals of the Missouri Botanical Gardens*, **79**, 474–499.

Hueber, F. M. & Galtier, J. 2002. *Symplocopteris wyattii* n. gen. et n. sp.: a zygopterid fern with a false trunk from the Tournaisian (Lower Carboniferous) of Queensland, Australia. *Review of Palaeobotany and Palynology*, **119**, 241–273.

Hughes, N. F. 1976. *Paleobiology of Angiosperm Origins*. Cambridge: Cambridge University Press.

Hughes, N. F. 1994. *The Enigma of Angiosperm Origins*. Cambridge: Cambridge University Press.

Hutchinson, G. & Thomas, B. A. 1996. *Welsh Ferns, Clubmosses, Quillworts and Horsetails: A Descriptive Handbook*. Cardiff, UK: National Museums & Galleries of Wales.

Jansonius, J. & McGregor, D. C. (eds.) 1996. *Palynology: Principles and Applications*, in 3 volumes. Dallas, TX: American Association of Stratigraphic Palynologists' Foundation.

Jennings, J. J. 1972. A polyvinyl chloride peel technique for iron sulphide petrifactions. *Journal of Paleontology*, **46**, 70–71.

Jones, T. P. & Rowe, N. P. (eds.) 1999. *Fossil Plants and Spores: Modern Techniques*. London: The Geological Society.

Jordan, G. J., Macphail, M. K. & Hill, R. S. 1996. A fertile pinnule with spores of *Dicksonia* from Early Oligocene sediments in Tasmania. *Review of Palaeobotany and Palynology*, **92**, 245–252.

Kendall, M. W. 1952. Some conifers from the Jurassic of England. *Annals and Magazine of Natural History, Series 12*, **5**, 583–594.

Kenrick, P. 1994. Alternation of generations in land plants: new phylogenetic and palaeobotanical evidence. *Biological Review*, **69**, 293–330.

Kerp, J. H. F. 1988. Aspects of Permian palaeobotany and palynology. X. The west- and central-European species of the genus *Autunia* Krasser emend. Kerp (Peltaspermaceae) and the form-genus *Rhachiphyllum* Kerp (callipterid foliage). *Review of Palaeobotany and Palynology*, **54**, 249–360.

Kerp, J. H. F. 1991. The study of fossil gymnosperms by means of cuticular analysis. *Palaios*, **5**, 548–569.

Kerp, J. H. F. 1996. Post-Varsican late Palaeozoic northern hemisphere gymnosperms: the onset to the Mesozoic. *Review of Palaeobotany and Palynology*, **90**, 263–285.

Kerp, J. H. F. & Barthel, M. 1993. Problems of cuticular analysis of pteridosperms. *Review of Palaeobotany and Palynology*, **78**, 1–18.

Kerp, J. H. F. & Haubold, H. 1988. Aspects of Permian palaeobotany and palynology. VIII. On the reclassification of the west- and central European species of the form-genus *Callipteris* Brongniart 1849. *Review of Palaeobotany and Palynology*, **54**, 135–150.

Kidston, R. & Lang, W. H. 1917–1921. On Old Red Sandstone plants showing structure, from the Rhynie Chert Bed, Aberdeenshire. *Transactions of the Royal Society of Edinburgh*, **51**, 761–784 (part 1); **52**, 603–627, 643–680, 831–854, 855–902 (parts 2–5).

Knowlton, F. H. 1913. The fossil forests of Arizona. *American Forestry*, **19**, 202–218.

Knowlton, F. H. 1917. A fossil flora from the Frontier Formation of southwest Wyoming. *U.S. Geological Survey Professional Paper*, **108-F**, 73–94.

Konijnenburg-van Cittert, J. H. A. van & Morgans, H. S. 1999. *The Jurassic Flora of Yorkshire*, Field Guide to Fossils. No. 8. London: Palaeontological Association.

Krings, M. 2000. The use of biological stains in the analysis of late Palaeozoic pteridosperm cuticles. *Review of Palaeobotany and Palynology*, **108**, 143–150.

Kvaček, Z. & Manum, S. V. 1993. Ferns in the Spitzbergen Palaeogene. *Palaeonotographica, Abteilung B*, **230**, 169–181.

Kvaček, Z. & Walther, H. 1995. The Oligocene volcanic flora of Suletice-Berand near Ústí Nad Labem, North Bohemia – a review. *Acta Musei Nationalis Pragae, B, Historia Nataturalis*, **50**, 25–54.

Labandiera, C. C., Dilcher, D. L., Davies, D. R. & Wagner, D. L. 1994. Ninety-seven million years of angiosperm-insect association: paleobiological insights into the meaning of coevolution. *Proceedings of the National Academy of Sciences, USA*, **91**, 122278–122282.

Lacey, W. S. 1963. Palaeobotanical techniques. In J. D. Carthy & C. L. Duddington, eds., *Viewpoints in Biology, Volume 2*. London: Butterworth, pp. 202–243.

Lang, P. J., Scott, A. C. & Stephenson, J. 1995. Evidence of plant-arthropod interactions from the Eocene

Branksome Sand Formation, Bournemouth, England: Introduction and description of leaf mines. *Tertiary Research*, **15**, 145–174.

Large, M. F. & Braggins, J. E. 2001. *Tree Ferns*. Portland, Oregon: Timber Press.

Long, A. G. 1960. On the structure of *Calymmatotheca kidstoni* Calder (emended) and *Genomosperma latens* gen. et sp. nov. from the Calciferous Sandstone Series of Berwickshire. *Transactions of the Royal Society of Edinburgh*, **64**, 29–44.

Long, A. G. 1996. *Hitherto*. Edinburgh: The Pentland Press.

Lyons, P. C., Morey, E. D. & Wagner, R. H. (eds.) 1995. *Historical Perspective of Early Twentieth Century Carboniferous Palaeobotany in North America*. Boulder, CO: Geological Society of America.

Macgregor, M. & Walton, J. 1955. *The Story of Fossil Grove*. Glasgow, UK: Public Parks Department.

Mamay, S. H. & Bateman, R. M. 1991. *Archaeocalamites lazarii*, sp. nov.: the range of Archaeocalamitaceae extended from the lowermost Pennsylvanian to the mid-Lower Permian. *American Journal of Botany*, **78**, 489–496.

Manchester, S. R. 1987. The fossil history of the Juglandaceae. *Monographs in Systematic Botany, Missouri Botanic Garden*, **21**, 1–137.

Manchester, S. R. 1992. Flowers, fruits, and pollen of *Florissantia*, an extinct malvalean genus from the Eocene and Oligocene of North America. *American Journal of Botany*, **79**, 996–1008.

Manchester, S. R. & Crane, P. R. 1983. Attached leaves, inflorescences, and fruits of *Fagopsis*, an extinct genus of Fagaceous affinity from the Oligocene Florissant Flora of Colorado, U.S.A. *American Journal of Botany*, **70**, 1147–1164.

Manchester, S. R. & Zavada, M. S. 1987. *Lygodium* foliage with intact sorophores from the Eocene of Wyoming. *Botanical Gazette*, **148**, 392–399.

Mapes, G. & Rothwell, G. W. 1991. Structure and relationships of primitive conifers. *Neues Jahrbuch für Geologie und Paläontologie, Abhandlungen*, **183**, 269–287.

Meyen, S. V. 1987. *Fundamentals of Palaeobotany*. London: Chapman and Hall.

Meyen, S. V. 1997. Permian conifers of Western Angaraland. *Review of Palaeobotany and Palynology*, **96**, 351–447.

Millay, M. A. 1997. A review of permineralised Euramerican Carboniferous tree ferns. *Review of Palaeobotany and Palynology*, **95**, 191–209.

Miller, C. N. 1971. Evolution of the fern family Osmundaceae based on anatomical studies. *Contributions from the Museum of Paleontology, The University of Michigan*, **23**, 105–169.

Miller, C. N. 1977. Mesozoic conifers. *Botanical Review*, **43**, 218–271.

Miller, C. N. 1982. Current status of Paleozoic and Mesozoic conifers. *Review of Palaeobotany and Palynology*, **37**, 99–114.

Niklas, K. J. & Banks, H. P. 1990. A reevaluation of the Zosterophyllophytina with comments on the origin of lycopods. *American Journal of Botany*, **77**, 274–283.

Pardoe, H. S. & Thomas, B. A. 1992. *Snowdon's Plants Since the Glaciers: A Vegetation History*. Cardiff, Wales: National Museum of Wales.

Phillips, T. L. & DiMichele, W. A. 1992. Comparative ecology and life-history biology of arborescent lycopsids in late Carboniferous swamps of Euramerica. *Annals of the Missouri Botanic Garden*, **79**, 560–588.

Pigg, K. B. & Rothwell, G. R. 2001. Anatomically preserved *Woodwardia virginica* (Blechnaceae) and a new filicalean fern from the Middle Miocene Yakima Canyon flora of central Washington, USA. *American Journal of Botany*, **88**, 777–787.

Pigg, K. B. & Trivett, M. L. 1994. Evolution of the glossopterid gymnosperms from Permian Gondwana. *Journal of Plant Research*, **107**, 461–477.

Poinar, G., Jr. 2002. Fossil palm flowers in Dominican and Mexican amber. *Botanical Journal of the Linnean Society*, **138**, 57–61.

Poinar, H. N., Cano, R. J. & Poinar, G. O. 1993. DNA from an extinct plant. *Nature*, **363**, 677.

Poort, R. J. & Kerp, J. H. F. 1990. Aspects of Permian palaeobotany and palynology XI. On the recognition of true peltasperms in the Upper Permian of western and central Europe and a reclassification of species formerly included in *Peltaspermum*. *Review of Palaeobotany and Palynology*, **63**, 197–225.

Rayner, R. J. 1992. *Phyllotheca*: the pastures of the Late Permian. *Palaeogeography, Palaeoclimatology, Palaeoecology*, **92**, 31–40.

Read, C. B. & Brown, R. W. 1937. American Cretaceous ferns of the genus *Tempskya*. *United States Geological Survey, Professional Paper*, **186-F**, 105–129.

Remy, W., Gensel, P. G. & Hass, H. 1993. The gametophyte generation of some Early Devonian land plants. *International Journal of Plant Science*, **154**, 35–58.

Retallack, G. J. & Dilcher, D. L. 1981. Early angiosperm reproduction: *Prisca reynoldsii*, gen. et sp. nov. from Mid-Cretaceous coastal deposits in Kansas, U.S.A. *Palaeontographica, Abteilung B*, **179**, 103–137.

Retallack, G. J. & Dilcher, D. L. 1988. Reconstructions of selected seed ferns. *Annals of the Missouri Botanical Garden*, **75**, 1010–1057.

Rex, G. M. 1983. The compression state of preservation of Carboniferous lepidodendrid leaves. *Review of Palaeobotany and Palynology*, **39**, 65–85.

Rex, G. M. & Chaloner, W.G. 1983. The experimental formation of plant compression fossils. *Palaeontology*, **26**, 231–252.

Robinson, S. R. & Miller, C. N. 1975. Glycol methacrylate as an embedding medium for lignitic plant fossils. *Journal of Palaeontology*, **49**, 559–561.

Roselt, G. von, Kupetz, M. & Beuge, P. 1982. Tertiäre Palmenwürzeln aus einer Tongrube bei Bad Freienwalde. *Abhandlungen des Staatlichen Museums für Mineralogie und Geologie zu Dresden*, **31**, 133–140.

Rössler, R. & Galtier, J. 2002a. First *Grammatopteris* tree ferns from the southern hemisphere – new insights in the evolution of the Osmundaceae from the Permian of Brazil. *Review of Paleobotany and Palynology*, **121**, 205–230.

Rössler, R. & Galtier, J. 2002b. *Dernbachia brasiliensis* gen. nov. et sp. nov.: a new small tree fern from the Permian of NE Brazil. *Review of Paleobotany and Palynology*, **122**, 219–263.

Rothwell, G. W. 1981. The Callistophytales (Pteridospermopsida). Reproductively sophisticated gymnosperms. *Review of Palaeobotany and Palynology*, **32**, 103–121.

Rothwell, G. W. 1984. The apex of *Stigmaria* (Lycopsida), rooting organ of Lepidodendrales. *American Journal of Botany*, **71**, 1031–1034.

Rothwell, G. W. 1996. Pteridophytic evolution: an often underappreciated phytological success story. *Review of Palaeobotany and Palynology*, **90**, 209–222.

Rothwell, G. W., Scheckler, S. E. & Gillespie, W. H. 1989. *Elkinsia* gen. nov., a Late Devonian gymnosperm with cupulate ovules. *Botanical Gazette*, **150**, 170–189.

Rothwell, G. W. & Stockey, R. A. 1989. Fossil Ophioglossales in the Paleocene of Western North America. *American Journal of Botany*, **76**, 637–644.

Rothwell, G. W. & Stockey, R. A. 1991. *Onoclea sensibilis* in the Paleocene of North America, a dramatic example of structural and ecological stasis. *Review of Palaeobotany and Palynology*, **709**, 113–124.

Rothwell, G. W. & Stockey, R. A. 1994. The role of *Hydropteris pinnata* gen. et sp. nov. in reconstructing the cladistics of heterosporous ferns. *American Journal of Botany*, **81**, 479–492.

Schopf, J. M. 1975. Modes of fossil preservation. *Review of Palaeobotany and Palynology*, **20**, 27–53.

Serbet, R. & Rothwell, G. W. 1992. Characterizing the most primitive seed ferns. I. A reconstruction of *Elkinsia polymorpha*. *International Journal of Plant Science*, **153**, 602–621.

Seward, A. C. 1914. *British Antarctic ('Terra Nova') Expedition, 1910. Natural History Report. Geology Vol. 1. No. 1, Antarctic fossil plants*. London: British Museum (Natural History).

Šimůnek, Z. 2007. New classification of the genus *Cordaites* from the Carboniferous and Permian of the Bohemian Massif, based on cuticle micromorphology. *Sborník Národního Muzea v Praze, Serie B, Přírodní Vědy*, **62**, 97–210.

Skog, J. E. & Dilcher, D. E. 1992. A new species of *Marsilea* from the Dakota Formation in Central Kansas. *American Journal of Botany*, **79**, 982–988.

Stace, C. A. 1989. *Plant Taxonomy and Biosystematics*, 2nd edn. London: Edward Arnold.

Stearn, W. T. 1992. *Botanical Latin*, 4th edn. Newton Abbot, UK: David and Charles.

Stein, W. E., Wight, D. C. & Beck, C. B. 1982. Techniques for preparation of pyrite and limonite permineralizations. *Review of Palaeobotany and Palynology*, **36**, 185–194.

Stewart, W. N. & Rothwell, G. W. 1993. *Paleobotany and the Evolution of Plants*, 2nd edn. Cambridge: Cambridge University Press.

Stockey, R. A. 1978. Reproductive biology of Cerro Cuadrado fossil conifers: ontogeny and reproductive strategies in *Araucaria mirabilis* (Spegazzini) Winhausen. *Palaeontographica Abteilung B*, **166**, 1–15.

Stockey, R. A. 1987. A permineralised flower from the Middle Eocene of British Columbia. *American Journal of Botany*, **74**, 1878–1887.

Stockey, R. A. 1988. Antarctic and Gondwana conifers. In T. N. Taylor & E. L. Taylor, eds., *Antarctic Paleobiology*. New York: Springer, pp. 179–191.

Stopes, M. C. 1918. *Married Love*. London: A. C. Fifield.

Stopes, M. C. 1919. On the four visible ingredients in banded bituminous coal. *Proceedings of the Royal Society B* **208**, 389–440.

Stopes, M. C. & Wheeler, R. V. 1919. *The Constitution of Coal*. London: Department of Scientific and Industrial Research, HMSO.

Stur, D. R. J. 1875. Beiträge zur Kenntniss der Flora der Vorwelt – Die Culm-Flora der Ostrauer und Waldenburger Schichten. *Abhandlungen Kaiserlich-königlichen Geologischen Reichsantalt*, **8**, 1–366.

Taylor, D. W. & Hickey, L. J. (eds.) 1996. *Flowering Plant Origin, Evolution and Phylogeny*. New York: Chapman and Hall.

Taylor, T. N. 1965. Paleozoic seed studies: a monograph of the American species of *Pachytesta*. *Palaeontographica, Abteilung B*, **117**, 1–46.

Taylor, T. N. & Millay, M. A. 1979. Pollination biology and reproduction in early seed plants. *Review of Palaeobotany and Palynology*, **27**, 329–355.

Taylor, T. N. & Millay, M. A. 1981. Morphologic variability of Pennsylvanian lyginopterid seed ferns. *Review of Palaeobotany and Palynology*, **32**, 27–62.

Taylor, T. N. & Smoot, E. L. (eds.) 1984. *Benchmark Papers in Systematic and Evolutionary Biology. 7. Paleobotany*, 2 vols. New York: Van Nostrand Reinhold.

Taylor, T. N. & Taylor, E. L. 1993. *The Biology and Evolution of Fossil Plants*, 2nd edn. Englewood Cliffs NJ: Prentice Hall.

Thomas, B. A. 1966. The cuticle of the Lepidodendroid stem. *New Phytologist*, **65**, 296–303.

Thomas, B. A. 1967. *Ulodendron*: Lindley and Hutton and its cuticle. *Annals of Botany New Series*, **31**, 775–782.

Thomas, B. A. 1970. A new specimen of *Lepidostrobus binneyanus* from the Westphalian B of Yorkshire. *Pollen et Spores*, **12**, 217–234.

Thomas, B. A. 1974. The Lepidodendroid stoma. *Palaeontology*, **17**, 525–539.

Thomas, B. A. 1977. Epidermal studies in the interpretation of *Lepidophloios* species. *Palaeontology*, **20**, 273–293.

Thomas, B. A. 1978. Carboniferous Lepidodendraceae and Lepidocarpaceae. *The Botanical Review*, **44**, 321–364.

Thomas, B. A. 1981. Structural adaptations shown by the Lepidocarpaceae. *Review of Palaeobotany and Palynology*, **32**, 377–388.

Thomas, B. A. 1986. *In Search of Fossil Plants: The Life and Work of David Davies (Gilfach Goch)*, Geology Series No. 8, Cardiff, UK: National Museum of Wales.

Thomas, B. A. 1997. Upper Carboniferous herbaceous lycopsids. *Review of Palaeobotany and Palynology*, **95**, 129–153.

Thomas, B. A. 2005. A reinvestigation of *Selaginella* species from the Asturian of the Zwickau coalfield, Germany and their assignment to the new subgenus *Hexaphyllum. Zeitschrift der Deutschen Gesellschaft für Geowissenschaften*, **156**, 1–12.

Thomas, B. A. & Cleal, C. J. 1993. *The Coal Measures Forests*. Cardiff, UK: National Museum of Wales.

Thomas, B. A. & Cleal, C. J. 1998. *Food of the Dinosaurs*. Cardiff, UK: National Museums & Galleries of Wales.

Thomas, B. A. & Cleal, C. J. 2000. *Invasion of the Land*. Cardiff, UK: National Museums & Galleries of Wales.

Thomas, B. A., Cleal, C. J. & Barthel, M. 2004. Palaeobotanical applications of incident-light dark-field microscopy. *Palaeontology*, **47**, 1641–1645.

Thomas, B. A. & Masarati, D. L. 1982. Cuticular and epidermal studies in fossil and living lycophytes. In D. F. Cutler, K. L. Alvin & C. E. Price, eds., *The Plant Cuticle*. Linnean Society Symposium Series, 10, pp. 363–378.

Thomas, B. A. & Meyen, S. V. 1984. A system of form-genera for the Upper Palaeozoic lepidophyte stems represented by compression-impression material. *Review of Palaeobotany and Palynology*, **41**, 273–281.

Thomas, B. A. & Spicer, R. A. 1987. *The Evolution and Palaeobiology of Land Plants*. London: Croom Helm.

Tidwell, W. D. & Ash, S. R. 1994. A review of selected Triassic to early Cretaceous ferns. *Journal of Plant Research*, **107**, 417–442.

Tidwell, W. D. & Nambudiri, E. M. V. 1989. *Tomlinsonia thomassonii*, gen. et sp. nov., a permineralized grass from the Upper Miocene Ricardo Formation, California. *Review of Palaeobotany and Palynology*, **60**, 165–177.

Tidwell, W. D. & Parker, L. R. 1990. *Protoyucca shadishi* gen. et sp. nov., an arborescent monocotyledon with secondary growth from the Middle Miocene of northwestern Nevada, U.S.A. *Review of Palaeobotany and Palynology*, **62**, 79–95.

Torsvik, T. H., Carlos, D., Mosar, J., Cocks, L. R. M. & Malme, T. 2002. Global reconstructions and North Atlantic palaeogeography 400 Ma to Recent. In E. A. Eide, ed., *BATLAS – Mid Norway Plate Reconstructions Atlas with Global and Atlantic Perspectives*. Oslo, Norway: Geological Survey of Norway, pp. 18–39.

Tralau, H. 1974. *Bibliography and Index to Palaeobotany and Palynology 1950–1970*, 2 vols. Stockholm, Sweden: H. Tralau.

Tralau, H. & Lundblad, B. (eds.) 1983. *Bibliography and Index to Palaeobotany and Palynology 1971–1975*, 2 vols. Stockholm, Sweden: Swedish Museum of Natural History and Swedish Natural Science Research Council.

Traverse, A. 2007. *Paleopalynology*, 2nd edn., *Topics in Geobiology 28*. Berlin: Springer.

Trivett, M. L. & Rothwell, G. W. 1991. Diversity among Paleozoic Cordaitales. *Neues Jahrbuch für Geologie und Paläontologie, Abhandlungen*, **183**, 289–305.

Vakhrameev, V. A., Dobruskina, I. A., Meyen, S. V. & Zaklinssskaja, E. D. 1978. *Paläozoische und mesozoische Floren Eurasiens und die Phytogeographie dieser Zeit.* Jena, Germany: G. Fischer.

Wagner, R. H. 1968. Upper Westphalian and Stephanian species of *Alethopteris* from Europe, Asia Minor and North America. *Mededelingen van de Rijks Geologische Dienst, Serie C, III-1*, **6**, 1–188.

Wagner, R. H. 1984. Megafloral zones of the Carboniferous. *Compte Rendu 9e Congrès International de*

Stratigraphie et de Géologie du Carbonifère (Washington, 1979), **2**, 109–134.

Wagner, R. H., Delcambre-Brousmiche, C. & Coquel, R. 2003. Una Pompeya Paleobotánica: historia de una marisma carbonífera sepultada por cenizas volcánicas. In R. Nuche, ed., *Patrimonio Geológico de Castilla-La Mancha*. Madrid: Enresa, pp. 448–477.

Walton, J. 1936. On the factors which influence the external form of fossil plants; with descriptions of the foliage of some species of the Palaeozoic equisetalean genus *Annularia* Sternberg. *Philosophical Transactions of the Royal Society of London, Series B*, **226**, 219–237.

Ward, L. F. 1885. Sketch of palaeobotany. *Report of the U. S. Geological Survey*, **5**, 363–469.

Watson, J. & Sincock, C. A. 1992. *Bennettitales of the English Wealden*. London: Palaeontographical Society.

Wellman, C. H. & Gray, J. 2000. The microfossil record of early land plants. *Philosophical Transactions of the Royal Society of London, Series B*, **355**, 717–732.

Wheeler, E. A., Baas, P. & Gasson, P. E. 1989. *IAWA List of Microscopic Features for Hardwood Identification*. Rijksherbarium, Leiden: IAWA Bulletin n.s., **10.3**.

White, M. E. 1986. *The Greening of Gondwana*. Frenchs Forest NSW, Australia: Reed.

Williamson, W. C. 1896. *Reminiscences of a Yorkshire Naturalist*, reprinted with additions by J. Watson & B. A. Thomas 1985. London: George Redway.

Wilson, S. & Yates, P. J. 1953. On two Dicksoniaceous ferns from the Yorkshire Jurassic. *Annals and Magazine of Natural History, Series 12*, **6**, 929–937.

Wittry, J. 2006. *The Mazon Creek Fossil Flora*. Doowners Grove IL: Earth Science Club of Northern Illinois.

Wolberg, D. & Reinhard, P. 1997. *Collecting the Natural World: Legal Requirements and Presumed Liability for Collecting Plants, Animals, Rocks, Minerals and Fossils*. Tucson, AZ: Geosciences Press.

Wolfe, J. A. 1997. Relations of environmental change to angiosperm evolution during the Late Cretaceous and Tertiary. In K. Iwatsuki & P. H. Raven, eds., *Evolution and Diversification of Land Plants*. Tokyo: Springer, pp. 269–290.

Zeiller, R. 1900. *Eléments de Paléobotanique*. Paris: Carré et Naud.

Zittel, K. von. 1901. *History of Geology and Palaeontology to the End of the Nineteenth Century*. London: W. Scott.

Zodrow, E. L., Cleal, C. J. & Thomas, B. A. 2001. *An Amateur's Guide to Coal: Plant Fossils on Cape Breton Island, Nova Scotia, Canada*. Sydney NS, Australia: University College of Cape Breton Press.

Index